Two Manuals Combined into One Book

Tactical Employment of Nonlethal Weapons

And

Security Force Assistance

Department of Defense

Pentagon
Publishing

PREFACE

1. Scope

This publication describes multi-Service tactics, techniques, and procedures (MTTP) for consideration and use during the tactical employment of nonlethal weapons (NLW) in support of warfighting personnel conducting training and tactical operations. This publication—

a. Provides an overview of NLW and its relationship to deadly force.

b. Provides NLW system description.

c. Describes the capability requirements of NLW.

d. Discusses fundamental concepts and training requirements involved with NLW.

e. Discusses the tactical employment considerations of NLW.

f. Discusses lessons learned from previous NLW use.

2. Purpose

a. This publication provides a single-source, consolidated reference on the tactical employment of NLW and supporting systems.

b. This publication is not intended to restrict the authority of the commander from organizing the force and executing the mission in a manner deemed appropriate.

3. Application

a. This publication provides commanders and their staffs unclassified guidance for NLW employment and planning. Commanders and staffs can use this publication to aid in the tactical employment of NLW during exercises and contingencies.

b. The United States (U.S.) Army, Marine Corps, Navy, Air Force, and Coast Guard approved this multi-Service publication for use.

4. Implementation Plan

Participating service command offices of primary responsibility will review this publication, validate the information and reference, and incorporate it in service manuals, regulations, and curricula follows.

Army. The Army will incorporate the procedures in this publication in U.S. Army training and doctrinal publications as directed by the Commander, U.S. Army Training and Doctrine Command (TRADOC). Distribution of this publication is in accordance with (IAW) initial distribution number (IDN) 115757.

Marine Corps. The Marine Corps will incorporate the procedures in this publication in U.S. Marine Corps doctrinal and training publications as directed by the Commanding General, U.S. Marine Corps Combat Development Command (MCCDC). Distribution of this publication is IAW Marine Corps Doctrinal Publication System (MCPDS).

Navy. The Navy will incorporate the procedures in this publication in U.S. Navy doctrinal and training publications as directed by the Commander, Navy Warfare Development Command (NWDC). Distribution of this publication is IAW Military Standard Requisition-

MARINE CORPS PCN: 143 000057 00
Air Force Distribution: F

ing and Issue Procedures (MILSTRIP) Desk Guide and Navy Standing Operating Procedure (NAVSOP) Pub 409.

Air Force. The Air Force will validate and incorporate appropriate procedures IAW applicable governing procedures. Distribution is IAW Air Force Instruction (AFI) 33-360.

Coast Guard. The Coast Guard will incorporate the procedures in this publication in U.S. Coast Guard doctrinal and training publications as directed by the Assistant Commandant for Operations, Director of Operations Policy, U.S. Coast Guard (USCG). Distribution of this publication is IAW Coast Guard standing operating procedures.

5. User Information

a. TRADOC, MCCDC, NWDC, Air Force Doctrine Center (AFDC), USCG, and the Air Land Sea Application (ALSA) Center developed this publication with the joint participation of the approving service commands. ALSA will review and update this publication as necessary.

b. This publication reflects current joint and service doctrine, command and control organizations, facilities, personnel, responsibilities, and procedures regarding NLW employment. Changes in service protocol, appropriately reflected in joint and service publications, will likewise be incorporated in revisions of this document.

c. We encourage recommended changes for improving this publication. Key comments to the specific page and paragraph and provide rationale for each recommendation directly to—

***FM 3-22.40** (FM 90-40)
***MCWP 3-15.8**
***NTTP 3-07.3.2**
AFTTP(I) 3-2.45
***USCG PUB 3-07.31**

FM 3-22.40 (FM 90-40) — **U.S. Army Training and Doctrine Command, Fort Monroe, Virginia**

MCWP 3-15.8 — **U.S. Marine Corps Combat Development Command, Quantico, Virginia**

NTTP 3-07.3.2 — **U.S. Navy Warfare Development Command, Newport, Rhode Island**

AFTTP(I) 3-2.45 — **Headquarters, Air Force Doctrine Center, Maxwell Air Force Base, Alabama**

USCG Pub 3-07.31 — **U.S. Coast Guard, Washington, D.C.**

15 January 2003

NLW

Tactical Employment of Nonlethal Weapons

TABLE OF CONTENTS

***This publication supersedes FM 90-40, MCRP 3-15.8, NWP 3-07.31, USCG Pub 3-07.31, dated 6 October 1998.**

FIGURES

Tables

EXECUTIVE SUMMARY

Tactical Employment of Nonlethal Weapons (NLW)

Today's world events mandate a need to project nonlethal force across all levels of war to enable our warfighters and leaders to effectively deal with a host of traditional as well as nontraditional threats. Now more than ever, the minimal level of tolerance for collateral damage and loss of human life, coupled with the tendency for the typical adversary to exploit the rules of engagement (ROE) to his benefit, necessitates an effective and flexible application of force through nonlethal weapons. Much like a rheostat switch in which power can be dialed up or down as desired, NLW provide tools to allow a commander to employ sufficient force to accomplish an objective without requiring the destruction of an enemy or the habitat. The intent of employing NLW is not to add another step in the progression of escalation with an adversary, but to add another tool to use anywhere along that continuum. This publication provides initial guidance for employing NLW in a tactical environment.

Overview of NLW

Chapter I describes the background, definition, and the concept description of NLW. This chapter discusses the development and employment of NLW, and explains the relationship of NLW to deadly force capabilities. The final portion of the chapter reviews policy per Office of Secretary of Defense instructions on employing NLW.

Purpose, System Description, and Organization

Chapter II describes the purpose of NLW and gives a general description of currently available commercial or government off-the-shelf equipment and munitions. The chapter also addresses organization structure, to include system location/distribution, personnel requirements, and mission requirements/tactical structure.

New Capability Requirements

Chapter III describes the applicability of NLW across the range of military operations. Additionally, this chapter discusses the core capabilities as stated in the Joint Mission Area Analysis (JMAA) as described in the Department of Defense Joint Concept for NLW.

Training With the NLW Capability Set

Chapter IV describes training requirements, topics, methods, locations, and a program of instruction from the Interservice Nonlethal Individual Weapons Instructor Course currently available for NLW options. This chapter also addresses additional training available from civilian law enforcement agencies.

Employment of NLW Capabilities

Chapter V establishes the fundamental understanding of the force continuum needed to employ NLW. It outlines employment prerequisites, mission-planning factors, and personnel requirements needed to employ NLW options. It further delineates NLW tasks according to the JMAA and provides considerations for preparation for unit tactical operations. This chapter also discusses employment objectives for NLW and ROE development. This chapter ends with multiple examples of service-specific crowd control formations used by units equipped with NLW.

Lessons Learned

Chapter VI focuses on NLW lessons learned from recent military operations. The chapter discusses how nonlethal technology has changed the way forces are looking at their adversaries in using NLW. It also covers the use of NLW along the force continuum in conjunction with ROE and the media.

PROGRAM PARTICIPANTS

The following commands and agencies participated in developing this publication:

Joint

U.S. Joint Forces Command, Suffolk, VA
U.S. European Command, J3/J5, APO AE
U.S. Pacific Command, J3832, Camp H. M. Smith, HI
U.S. Central Command, J3, MacDill AFB, FL
U.S. Southern Command, SCJ5-PS, Miami, FL
U.S. Space Command, SPJ5X, Peterson AFB, CO
U.S. Special Operations Command, HQ USSOCOM, SOOP-PTE, MacDill AFB, FL
Joint Nonlethal Weapons Directorate, Quantico, VA
Interservice Nonlethal Individual Weapons Instructor Course, Fort Leonard Wood, MO

Army

Headquarters, Department of the Army, ODCSOPS (DAMO-FDT), Pentagon, Washington, D.C.
U.S. Army Training and Doctrine Command, DCSDCS (ATDO-A), Fort Monroe, VA
U.S. Army Training and Doctrine Command, DCSDEV (ATCD-SN), Fort Monroe, VA
U.S. Army Infantry Center, DCD, Fort Benning, GA
U.S. Army Military Police School, ATTN: DOTD, MANSCEN (ATZT-DT-WM-M), Fort Leonard Wood, MO
USAJFKSWCS, DOTD, Fort Bragg, NC
National Ground Intelligence Center, Charlottesville, VA

Marine Corps

Marine Corps Combat Development Command, Quantico, VA
MARCORSYSCOM, Quantico, VA
USMC Warfighting Lab, Quantico, VA
MCSFT Company, MCSF Battalion, Chesapeake, VA
2d FAST MCSF Battalion, Yorktown, VA
1st MARDIV Military Police Company, HQ Battalion, Camp Pendleton, CA
Marine Corps Detachment Military Police and Chemical School, Fort Leonard Wood, MO
Special Operations Training Group, I Marine Expeditionary Force, Camp Pendleton, CA
Special Operations Training Group, II Marine Expeditionary Force, Camp Lejeune, NC

Navy

Commander, Navy Warfare Development Command (NWDC), Newport, RI
USN Surface Warfare Center, Crane, IN
USN Surface Warfare Center, Dahlgren, VA

Air Force

AFDC/DJ, Langley AFB, VA
Headquarters, ACC/SFOT, Langley AFB, VA
Air Force Security Forces Center/SFOR, Lackland, AFB, TX
Headquarters, USAF/XOF, Washington, D.C.

Coast Guard

Commandant, U.S. Coast Guard, (G-OPL-2), Washington, D.C.

Chapter I

Overview of Nonlethal Weapons

1. Background

a. In today's operational environment, United States (U.S.) forces regularly perform missions that were almost inconceivable a few decades ago. In this new environment, lethal firepower or the threat of its use may no longer be the default solution to all crises or problems. Senior leaders face a new level of public and media sensitivity and scrutiny concerning the proper role of the military as an instrument of national power. Field commanders must understand these sensitivities and attempt to achieve an appropriate military force. Junior leaders must apply the resulting decisions wisely, often in changing situations filled with uncertainty and danger.

b. Throughout history, changes in culture and technology have influenced the character of military force and the manner in which it is employed. These changes are an attempt to maximize military force in the force continuum. In the complex and changing political and social domain of the modern world, nonlethal capabilities may offer the opportunity to increase or maximize the military as an instrument of national power.

c. Nonlethal weapons (NLW) must now be considered an integral part of joint operational capabilities. Joint forces anticipate that NLW technology will advance and provide additional options for force commanders.

2. Definition

a. Department of Defense Directive (DODD) 3000.3, Policy for Nonlethal Weapons, 9 July 1996, defines NLW as "weapons that are explicitly designed and primarily employed so as to incapacitate personnel or materiel while minimizing fatalities, permanent injury to personnel, and undesired damage to property and the environment." Furthermore, "unlike conventional lethal weapons that destroy their targets principally through blast, penetration, and fragmentation, nonlethal weapons employ means other than gross physical destruction to prevent the target from functioning. Nonlethal weapons are intended to have relatively reversible effects on personnel and materiel."

b. The term "nonlethal" does not mean zero mortality or nonpermanent damage; these are goals and not guarantees of these weapons. NLW can add flexibility to combat operations and enhance force protection by providing an environment in which friendly troops can engage threatening targets with reduced risk of noncombatant casualties and collateral damage.

3. Concept Description

Nonlethal capabilities within the context of the service capability sets (see Appendix E) may provide armed forces additional offensive and defensive tools to assist in the incapacitation and destruction of an enemy, capture of important intelligence sources, crowd control, civil disturbances, enforcing mandates, and force protection in accordance with the specified rules of engagement (ROE) for a specific operation. These operations include small-scale contingencies (SSC), military operations other than war (MOOTW), stability and support operations (SASO), and civil disturbances. The U.S. Army's Task Force (TF) FALCON in Kosovo and U.S. armed forces located at Camp X-RAY in Guantanamo Bay, Cuba are superb examples of the use of NLW to prevent escalation within an operation.

4. Background on Development and Employment

a. Military forces have long used nonlethal force to influence behavior of people and nations, defeat adversaries with minimum use of lethal capabilities, and weaken adversaries to defeat them with conventional arms. Examples of nonlethal means include a show of force, physical obstacles, riot control agents (RCA), noise to create or enhance psychological effects, smoke and obscurants to mask operations or defeat homing and guidance mechanisms, and using light to disorient combatants. These nonlethal means will remain relevant in future operations and are a part of an evolutionary process of weapons development. Joint operations in Somalia provided increased awareness of the need to incorporate nonlethal options into the fighting capabilities. Therefore, the Department of Defense (DOD) established the Joint Nonlethal Weapons Program in 1996 to research, develop, and acquire these capabilities.

b. Increased interaction between U.S. forces and civilian populations has become a feature of the contemporary operational landscape that is likely to remain the case in the future. Three factors account for this forecast:

(1) Worldwide patterns of population growth and migration have resulted in an expansion of urban culture within established industrialized nations, as well as many pre-industrial and emerging third world societies (60 percent of the world's population live in built-up areas). The prevalence of urbanization in many crisis-prone regions of the world creates the potential for large, vulnerable groups of noncombatants to be entrapped in volatile confrontations involving the deployment of military forces in military operations in urban terrain (MOUT).

(2) U.S. forces increasingly operate in the challenging environment known as MOOTW; a category that includes such missions as humanitarian assistance, disaster relief, noncombatant evacuations, and the various types of peace operations. These operations commonly involve close and continual interaction between U.S. forces and noncombatant civilians. Some MOOTW scenarios include the presence of paramilitary forces, armed factions, or rogue elements that present a constant, but uncertain threat to joint forces. In these situations, the mission of military forces includes preventive tasks. In other words, U.S. forces enhance mission accomplishment by preventing individuals or factions from carrying out specified undesirable activities, such as rioting, looting, attacking, harassing, or threatening. Often, the adversary blends in with the local population of innocent citizens. At other times, sectors of the local population may rise against U.S. forces and become active participants in acts of violence. Factional alignments, the level of violence, and the threat to mission accomplishment, may change frequently with little or no warning. Under such circumstances, the identity of adversaries is uncertain and the use of deadly force for purposes other than self-defense may be constrained by ROE or the judgment of the on-scene commander.

(3) From a defense-in-depth or situational understanding perspective, the force protection, homeland security, and the antiterrorism missions, require that forces assess the use of NLW as a response to potential asymmetric attacks. Asymmetric threats run the spectrum of adversarial use of innovative or nontraditional strategies, tactics, or technologies, to exploit vulnerabilities and strike a decisive blow while avoiding the strengths of an opponent. The objectives of these strikes are intended to undermine national will. Terrorist attacks such as those directed against the USS Cole, World Trade Center, and Pentagon, highlight the severity of the asymmetric threat and its resultant effects.

5. Relationship of NLW to Deadly Force

a. NLW provide "the lesser means" that could be considered prior to employing deadly force. The services definition of deadly force is as follows:

(1) U.S. Army – "Force that a person uses causing, or that a person knows, or should know, will create a substantial risk of causing death or serious bodily harm." (Army Regulation (AR) 190–14, Military Police Carrying of Firearms and Use of Force for Law Enforcement and Security Duties, 12 March 1993).

(2) USMC – "That force which a person uses with the purpose of causing death or serious bodily harm or which a reasonable and prudent person would consider likely to create substantial risk of causing death or serious bodily harm. It will be used as a last resort when all lesser means have failed." Marine Corps Order (MCO) 5500.6F, Arming of Security and Law Enforcement Personnel and the Use of Force, 20 July 1995.

(3) USN – "Force that a person uses causing, or that a person knows or should know would create a substantial risk of causing death or serious bodily harm." Secretary of The Navy Instruction (SECNAVINST) 5500.29B, Use of Deadly Force and the Carrying of Firearms by Personnel of the Department of the Navy in Conjunction with Law Enforcement, Security Duties and Personal Protection, 29 September 1992.

(4) USAF – "Force that causes, or that you know has a substantial risk of causing death or serious bodily harm." Air Force Instruction (AFI) 31-207, Arming and Use of Force by Air Force Personnel, 1 September 1999.

(5) USCG – "Deadly force is any force that is likely to cause death or other serious physical injury." (COMDTINST M16247.1B, Maritime Law Enforcement Manual, 9 August 2001.)

b. The commitment of military power to resolve crises has traditionally involved the use of deadly force or the implicit or explicit threat of the use of deadly force. Military units are trained, organized, and equipped for this purpose. A force armed only with traditional military weapons normally has two options for enforcing compliance: threating deadly force and applying deadly force. This limitation creates a critical vulnerability that belligerents may quickly discern and use to their advantage.

c. Nonlethal capabilities provide a wider range of options that augment, but do not replace traditional means of deadly force. The option to resort to deadly force must always remain available when the commander believes it is appropriate to the mission. DODD 3000.3, states "the availability of NLW shall not limit a commander's inherent authority and obligation to use all necessary means available and to take all appropriate action in self-defense." The existence of nonlethal capabilities does not represent the potential for "nonlethal war," and unrealistic expectations to that effect must be vigorously avoided. Noncombatant casualties, to include serious injuries and fatalities, are often an unavoidable outcome when employing military power, regardless of NLW availability. Augmenting deadly force with NLW should be fundamental to planning and executing any operation where the commander is contemplating deadly force.

d. The commander must clearly articulate the roles of NLW within the ROE. NLW is an additional means of employing force throughout the force continuum for the particular purpose of reducing the probability of death or serious injury to noncombatants or belligerents. Using deadly force must always remain an inherent right of individuals in instances when they, their fellow servicemen, or personnel in their charge are threatened with death or serious bodily harm. NLW add flexibility to combat operations and enhance force protection by providing the field commander with options to engage threatening targets with limited risk of noncombatant casualties and collateral damage.

e. Commanders and public affairs officers (PAO) must prepare personnel to address media questions and concerns regarding the role of NLW. Operational experience indicates that new capabilities give rise to significant media interest. Personnel participating in interviews or briefings must be prepared to address the role of NLW. It must be clear that the presence of NLW in no way indicates abandoning the option to employ deadly force in appropriate circumstances.

6. Advantages of Employing NLW

NLW provide commanders the flexibility to influence the situation favorably with increased safety to friendly forces and reduced risk of both noncombatant fatalities and collateral damage. Some advantages follow:

a. NLW can be more humane and consistent with the political and social implications implicit in humanitarian and peacekeeping missions.

b. The force that properly employs nonlethal options can gain advantages over those who rely on lethal options alone. Because the degree of provocation required to employ NLW is substantially less than lethal options, the force can take a more proactive position that may diminish the likelihood of having a situation escalate to a point where deadly force is required to resolve a conflict.

c. NLW options can be less likely to provoke others. However, commanders should be cautioned using NLW does not preclude an undesired result.

d. Demonstrated restraint can greatly diminish feelings of anger and remorse when deadly force is required after nonlethal options fail.

e. Using NLW can facilitate post incident stabilization by reducing populace alienation and collateral damage.

f. NLW can reduce the possibility of injury to friendly forces compared to forces without NLW.

7. Policy Per Office of the Secretary of Defense Instructions on Employment of NLW

a. DODD 3000.3 states, "nonlethal weapons, doctrine, and concepts of operation shall be designed to reinforce deterrence and expand the range of options available to commanders. Nonlethal weapons should enhance the capability of U.S. Forces to accomplish the following objectives:

(1) Discourage, delay, or prevent hostile actions.

(2) Limit escalation.

(3) Take military action in situations where use of lethal force is not the preferred option.

(4) Better protect our forces.

(5) Temporarily disable equipment, facilities, and personnel.

(6) Nonlethal weapons should also be designed to help decrease the post-conflict costs of reconstruction."

b. DODD 3000.3 continues by stating, "The availability of nonlethal weapons shall not limit a commander's inherent authority and obligation to use all necessary means available and to take all appropriate action in self-defense. Neither the presence nor the potential effect of nonlethal weapons shall constitute an obligation for their employment or a higher standard for employment of force than provided for by applicable law. In all cases, the United States retains the option for immediate use of lethal weapons, when appropriate, consistent with international law."

c. DODD 3000.3 further states, "Nonlethal weapons shall not be required to have a zero probability of producing fatalities or permanent injuries. However, while complete avoidance of these effects is not guaranteed or expected, when properly employed, nonlethal weapons should significantly reduce them as compared with physically destroying the same target." Nonlethal weapons may be used in conjunction with lethal weapon systems to enhance the latter's effectiveness and efficiency in military operations. This shall apply across the range of military operations to include those situations where overwhelming force is employed."

d. According to paragraph 3.21 of the Interium Guidance issued by the Deputy Secretary of Defense, 30 October 2002, all developmental weapon systems must be submitted for legal review and must receive favorable findings before they can be incorporated into the force. This review includes examination for consistency with applicable international and domestic laws, including the laws of war, various arms control treaties, and protocols. U.S. forces can be assured that any NLW they are issued have passed this critical test. Medical and policy reviews should also be considered prior to fielding any new weapon systems. Commanders and operators should consult their specific ROE for guidance.

e. Some capabilities, although designed to minimize fatalities and serious injuries, may have effects that could actually discourage their use. Therefore, the basic decision to employ military force in defense of national interests is usually a matter of intense public concern. The manner in which that force is exercised is subjected to the same scrutiny.

f. The presence of NLW on the battlefield does not mean they must be used. It also does not mean they must be used prior to the use of lethal force. In all cases, the U.S. retains the option for immediate use of lethal weapons when the on-scene commander deems it appropriate. Additionally, the immediate use of lethal weapons must be consistent with U.S. and international law and the ROE.

Chapter II

PURPOSE, SYSTEM DESCRIPTION, AND ORGANIZATION

1. Purpose

The purpose of NLW is to enhance the ability of military commanders to conduct operations across the spectrum of conflict. Over the past several years, military operations have gravitated towards the lower end of this spectrum and have involved U.S. forces in various missions. These missions include SSC, SASO, MOOTW, and civil disturbances where the ROE are inherently restrictive. NLW are designed to address the special requirements of low-intensity environments. They provide tools that enable the individual, team, or unit, to anticipate and/or respond to provocation with more appropriate means than through the use of deadly force. Essentially, NLW provide U.S. forces with a near-term capability for effectively controlling the nontraditional battlefield, within the constraints levied by ROE, by reducing the risk to friendly forces, mitigating casualties, and minimizing collateral damage.

2. System Description

a. NLW Capability Set. The services have, and are continuing to procure, versatile packages of NLW comprised of commercial off-the-shelf (COTS) and government off-the-shelf (GOTS) equipment and munitions. The government is also exploring advanced technologies for future procurements. In addition to addressing contingency requirements, the services' NLW capability sets provide limited sustainment training ammunition and equipment.

b. NLW Capability Set—Component Categories. Each service determines the contents of its own capability set. Examples of the services current configurations are shown in Appendix E. The NLW capability set components can be divided into four distinct categories: personnel protectors, personnel effectors, mission enhancers, and training devices. The following examples are not all-inclusive.

(1) Personnel Protectors. Personnel protectors include items such as face shields, riot shields, and shin guards that protect the individual from blunt trauma injuries inflicted by thrown objects, clubs, etc.

(2) Personnel Effectors. Personnel effectors include items such as riot batons, malodorants, stingball grenades, pepper sprays, and other kinetic rounds (e.g. sponge grenades) designed to discourage, disorient, or incapacitate individuals or groups.

(3) Mission Enhancers. Mission enhancers include items such as bullhorns, spotlights, caltrops, isotropic radiators (stun/flash), and barriers/entanglers. These items are designed to facilitate target identification and crowd control, and to limit personnel and vehicular movement.

(4) Training Devices. Training devices include items such as training suits, training rounds, training batons, and inert pepper sprays. They are designed to facilitate realistic hands-on scenario training in preparation for operations.

c. System Relationships and Inter/Intraoperability. The NLW capability set is designed to provide an operating force (ranging from squad through battalion-sized units [see Appendix E]) with tactically deployable nonlethal equipment. Fielding of this capability set enhances the operation of systems and equipment already fielded and planned within the joint community. The NLW capability set components are compatible with military equipment that is already fielded or planned for future fielding. The components of the NLW capability set are supported within current military logistic capabilities.

d. Augmentation of Organic Equipment. The operational intent of the capability set is to augment an operating force with the required nonlethal equipment to enhance its force projection capability. The NLW capability set is additional equipment that must be incorporated into lift and load planning.

3. Organization

Organization/organizational structure is not anticipated to change as a result of NLW capability sets being added to the inventory.

a. System Location/Distribution. The United States Army (USA), United States Marine Corps (USMC), and United States Navy (USN) have procured and fielded capability sets to the operating forces, security forces, bases, and stations. All services have current plans to field additional capability sets as required. At the time of this printing, the United States Air Force (USAF) has identified its capability sets to field in the near future. The United States Coast Guard (USCG) also intends to field NLW.

b. NLW Capability Set Composition. Appendix E describes the current and proposed capability sets.

c. Personnel Requirements. To date, no change or increase in military specialty codes have been determined in order to field, operate, and maintain the NLW capability set. Each service determines its own training personnel requirements.

d. Mission Requirements/Tactical Structure.

(1) The mission requirements and tactical structure pertinent to the employment of the NLW will vary depending on the type of contingency and subsequent task organization of the employing unit. Mission requirements, derived from commander's guidance, should be satisfied through the appropriate combination of lethal and nonlethal capabilities.

(2) NLW provide commanders with a means of shaping their area of operations (AO) short of employing deadly force. However, it is imperative that U.S. forces retain the means and ability to employ lethal force in the event that nonlethal force fails to achieve desired outcomes. For example, NLW employment may provoke a lethal response from the adversary. Without the imminent threat of lethal force, nonlethal munitions will prove to be less effective and possibly even counterproductive. More importantly, denying U.S. forces a lethal capability would place them at unacceptable risk. If nonlethal force is employed without the threat of lethal force, the belligerent may opt to exploit a perceived vulnerability. In this case, the use of NLW becomes counterproductive, because it serves only to aggravate the threat and causes the situation to deteriorate.

Chapter III

NLW CAPABILITY REQUIREMENTS

1. Applicability Across the Range of Military Operations

a. U.S. forces conduct military operations across a large spectrum. This spectrum spans from humanitarian assistance and peacekeeping to strategic nuclear exchange. Since the end of the Cold War, the U.S. military has been increasingly involved with operations at the lower end of the spectrum. These operations focus on deterring war, resolving conflict, supporting civil authorities and promoting peace. NLW may be useful across many categories of these operations. The range of military operations includes various levels of combat "intensity," and NLW provide one more force option to deal with the varying intensity. MOOTW may be found across the range of military operations and often include situations where noncombatants are mingled with the adversaries or involved in acts of violence, such as rioting or looting. In such circumstances, NLW provide commanders the flexibility to influence the situation favorably with reduced risk of noncombatant fatalities and collateral damage. NLW also provide a means by which to ascertain the intent of individuals/belligerents. Some characteristics of operations that present the opportunity to use NLW are—

(1) Close contact between friendly forces and noncombatant civilians.

(2) The presence of paramilitary or irregular forces.

(3) Armed civilian gangs, unlawful combatants, or militia.

b. These characteristics present a real but ill-defined threat. Under such circumstances, the commander's judgment or ROE may restrict using deadly force for other than self-defense. NLW may provide U.S. forces flexibility to resolve situations where the use of deadly force is restricted. These weapons provide flexibility by—

(1) Allowing U.S. forces to apply measured military force.

(2) Reducing risk of serious noncombatant causalities.

(3) Accomplishing the mission.

(4) Providing force protection.

(5) Retaining the initiative.

(6) Allowing U.S. forces to remain relevant and capable in increasingly complex and chaotic environments.

c. The need to reduce the risks of serious injury to personnel is not limited to crowd control scenarios. In MOUT, some of the local civilian populace may remain in an urban area in the midst of battle. The traditional solution to such challenges has been restricting ROE. Nonlethal capabilities may provide commanders the flexibility to adapt a fluid approach to MOUT, which allows their subordinates the freedom of action to employ appropriate levels of measured military force to accomplish their mission.

d. Another example of NLW applications in MOOTW is in peace enforcement. In these operations, NLW are used in an area denial role to separate belligerents without resorting to deadly force. They are also used to support humanitarian relief organizations in protecting food and medical distribution sites. The presence of NLW will not limit a commander's inherent authority and obligation to use all necessary means available and to take appropriate action in self-defense.

2. Core Capabilities

Core capabilities are those fundamental competencies that enable U.S. forces to achieve desired operational objectives. NLW provide a flexible means to protect friendly forces, high-value assets, and influence the actions of potential adversaries and/or noncombatants. These goals can be achieved without resorting to lethal force and in a manner that may minimize collateral damage. The core capabilities associated with nonlethal effects fall into three major categories as stated within the Joint Mission Area Analysis (JMAA) for Nonlethal Weapons (approved December 2000 by the Joint Requirements Oversight Council): counterpersonnel, countermateriel, and countercapability.

a. Counterpersonnel Capabilities. The DOD Joint Concept for Nonlethal Weapons concept paper, 5 January 1998, describes required nonlethal counterpersonnel capabilities for U.S. forces as follows:

(1) "The application of military force to accomplish a mission with reduced risk of fatalities or serious casualties among noncombatants—or even, in some instances, among enemy forces." There are several specific nonlethal counterpersonnel capabilities to be considered. These include, but are not limited to, the means to influence the behavior and activities of a potentially hostile crowd and the capability to bring a mob engaged in a riot under control. While there are many similarities in these two capabilities, each involves unique challenges, which may require radically different solutions.

(2) "Capability to incapacitate personnel." For the purposes of this document, "incapacitation" is achieved when the weapon's effects render personnel unable to physically or mentally (real or perceived) act in a hostile or threatening manner. In keeping with the guiding principles for NLW, this incapacitation should be reversible through the passage of time. The focus will be on capabilities that affect both individuals and groups of people.

(3) "Capability to deny personnel access to an area (land, sea, or air)." NLW can provide new possibilities for barrier planning in any type of military operation, including systems that produce physical or mental discomfort to those who enter the denied area.

(4) "Capability to clear facilities and structures of personnel." The Joint Concept for NLW states this capability "will facilitate MOUT by reducing the risks of noncombatant casualties and collateral damage while simultaneously minimizing the advantages accruing to an enemy defending a built up area." This capability is applicable to all clearing missions that may be conducted (e.g., Maritime Interception Operations [MIO]) and could assist in providing enhanced force protection.

(5) "Capability to seize personnel, which could include some combination of technologies inherent in other core capabilities, for example, incapacitation and entangling devices (which may aid in seizing personnel)." This capability is intended to augment lethal means used to capture specified individuals, such as enemy combatants or persons who are inciting a mob or crowd to violence.

b. Countermateriel Capabilities. The Joint Concept for Nonlethal Weapons describes countermateriel capabilities as follows:

(1) Rendering equipment and facilities unusable without complete destruction by attacking only weapons of war and supporting infrastructure while minimizing risk to noncombatants. This could enhance operations by defusing potentially volatile situations where conventional military means might otherwise prove counterproductive.

(2) Disabling or denying the use of vehicles, vessels, and aircraft entry into targeted areas or access within an AO. This capability may include physical barriers that impede or stop movement. Alternative systems may alter combustion properties of fuels

and viscosity of lubricants, embrittle or decay materials, attack tires, gaskets, or hoses, or act as adhesives to fuse metal parts. The requirement also includes the capability to facilitate boarding and clearing operations.

c. Counter Capability Capabilities. For U.S. forces counter capability capabilities are described within the JMAA for NLW as follows:

(1) "Disable or neutralize facilities and systems." This includes a nonlethal capability for disabling/neutralizing electrical generating facilities, command, control, communications, computers, intelligence, surveillance, and reconnaissance (C4ISR) systems, Integrated Air Defense Systems (IADS), weapons systems, optical sensors, electrical sensors, and navigation capabilities with such controlled effects as to allow for selective, precise engagement. Other needs include deceiving reconnaissance with illusions of barriers, obstacles, and/or forces, and concealing/covering friendly activities or movements, particularly within complex urban terrain.

(2) "Deny the use of weapons of mass destruction (WMD)." Mission needs include rendering a WMD inoperative; containing the potential release of deadly agents/contaminants; and preventing or neutralizing the production, storage, deployment (transport), employment, and delivery of WMD. Nonlethal capabilities in combating WMD should be designed to permit engagement within a populated or sensitive terrain, prior to the initiation of armed conflict.

3. Limitations

The limitations of NLW are effectiveness, perception, and legality.

a. The effectiveness of NLW is dependent on factors such as motivation, age, environmental and human factors, range of engagement, and numerous others. NLW can be lethal if improperly applied. Furthermore, countermeasures for thwarting virtually all nonlethal options are usually apparent, quickly learned, and readily available. Because they are not intended to kill, nonlethal options teach adversaries what to avoid in the future. Small unit commanders must prepare to keep one step ahead of belligerents.

b. The perceptual limitation is that personnel may misunderstand the appropriate applications of NLW across the range of military operations. The incorrect perception that NLW will allow wars and MOOTW to be prosecuted without casualties may lead to conflicting expectations between political and military leaders. These conflicts can result in U.S. forces being vulnerable to misplaced or inconsistent ROE. All leaders, political and military, involved in planning and executing military missions must understand there are no "nonlethal operations."

c. Some proposed NLW might be forbidden by law or policy. Accordingly, it is essential that all NLW developments be evaluated by appropriate authorities to ensure they comply with the law of war, U.S. law, and U.S. treaty obligations. Use of NLW containing chemical agents, for instance, must be evaluated in the context of the Chemical Weapons Convention. Also, using RCA in an armed conflict requires Presidential approval.

Chapter IV

TRAINING

1. Fundamental Concepts and Training Requirements

The successful accomplishment of any operation in which nonlethal measures are employed requires extensive preparation, of which individual, unit, medical support, and unit training are vital parts. Training should be designed to give individuals an understanding of the entire subject area and enable them to function efficiently as members of a unit. This training must be intensive and realistic. Training with the NLW capability set, and other NLW, is critical to employ these weapons with confidence. The training flow should be—train leaders, train instructors, train unit, exercise. Qualified instructors should conduct initial unit training. Unit sustainment training should be conducted per individual service and unit requirements.

2. Training of Instructors

a. Verbal communication skills, open-hand control techniques, RCA, effects of impact weapons, and civil disturbance tactics, techniques, and procedures (TTP) are areas trained to standard at the Interservice Nonlethal Individual Weapons Instructor Course (INIWIC) held at the Marine Corps Detachment, Fort Leonard Wood, Missouri. INIWIC is an Interservice Training Review Organization (ITRO) course, and quotas are available to any service through normal training reservation channels. A course outline is provided in Appendix C. This is the only formal DOD NLW instructor-training course. The training requirements outlined in the INIWIC represent the minimum training standards to include in developing NLW and tactics instructors. Additionally, the United States Army Military Police School (USAMPS) has developed a training support package that includes Army-specific TTP for small units (company level and below) in using NLW capabilities. The reference for this is the U.S. Army's website at: www.wood.army.mil.

b. The INIWIC is a "train-the-trainer" program to provide commanders NLW instructors within their command. Furthermore, the military services are not precluded from developing their own training based upon the INIWIC curriculum. The commander can then tailor the NLW instruction to fit mission needs. Establishing the service-level INIWIC serves as an assurance measure, giving the commander a quantifiable means of certifying his nonlethal instructors.

3. Nonlethal Training Topics for Deploying Units

Each service and unit will have unique training requirements based on mission type, unit size, deployment area, etc. As a minimum, the training instructor should cover the following topics in any training for units deploying with nonlethal capabilities. They are part of the INIWIC curriculum.

NOTE: U.S. Army training standards are contained in Army Regulation (AR) 350-38, Training Devices Policies and Management, 15 October 1993; USMC training standards for NLW are contained in MCO 1510.112, Individual Training Standards for Nonlethal Weapons, 27 May 1998.

---

a. Force Continuum. This is an introduction to the federal force continuum model as outlined by MCO 5500.6F, Arming of Law Enforcement and Security Personnel and the Use of Deadly Force; AR 190-14; AFI 31-207, Arming and Use of Force by Air Force Personnel (1 September 1999) (or other service orders). Identifying the proper levels of force, understanding levels of resistance, and how nonlethal technologies affect the force continuum are the minimum tópics the instructor should cover.

b. Crowd Dynamics/Crowd Control. This is an introduction to the differences between crowds, mobs, and riots. This course teaches the student basic crowd control techniques that will be easily applied to various situations. The student should be familiarized with tactics and techniques found in field manual (FM) 19-15, MCWP 3-11.1, Rifle Company, TC 90-1, and Air Force manual (AFMAN) 31-201V6, Civil Disturbance (1 April 2002), but will also consider nontraditional and small unit application.

c. Communication Skills. This is an introduction to the stages of conflict management, verbal aggression, nonverbal communication, physical aggression, physiological diversions, and proper mental conditioning. The first NLW skill an individual must have is the ability to communicate. General A. M. Gray, former Commandant of the Marine Corps, once said, "A warrior's most formidable weapon is his mind." The results of an individual's mental processes are evident through actions and words, and the words and communication skills used in dealing with people are truly the first line of nonlethal options.

d. Oleoresin Capsicum (OC) Aerosol and other RCA Training. This is an introduction to the uses of OC/pepper spray and other RCA. The student should have an appreciation for decontamination requirements, storage, and shelf life of OC/RCA. The instructor will also teach legal and tactical considerations. INIWIC certified instructors should teach existing training programs that include live agent training.

e. Open-Hand Control. This is an introduction to pressure point control techniques, unarmed self-defense measures, weapon retention techniques, and other submission/restraint/search techniques. The instructor should cover handcuffing/flex-cuffing for contact team members.

NOTE: USMC plans to replace the current open-hand defensive tactics training with the USMC Martial Arts Program.

---

f. Impact Weapons. This is an introduction to using the rigid straight baton, collapsible straight batons, side handle batons, and/or riot control batons.

g. ROE/Law of Armed Conflict (LOAC). This is an introduction to the ROE/LOAC and their relationship. (Refer to DODD 5100.77, AR 27-1, and NWP 1-14M).

h. Nonlethal Munitions—Employment and Tactics. This subcourse is an introduction to the capabilities and employment of available NLW. NLW provide additional unit force protection capabilities. Introduction to tactics is outlined in Chapter 5. Appendix D is a sample unit training schedule covering the topics listed above. U.S. Army Training and Doctrine Command (TRADOC) has developed training support packages that will include company level and below TTP.

4. Leader Training

a. An understanding of the topics covered in the INIWIC curriculum is essential. However, commanders/leaders should also possess an in-depth understanding of the ROE, their implementation, as well as the politics, culture, and demographics of the area to which they are deploying.

b. Commanders should involve their medical personnel, public affairs (PA) representatives, civil affairs (CA), psychological operations (PSYOP), staff judge advocate

(SJA), engineers, explosive ordnance disposal (EOD), and military police (MP) in training exercises when possible. Involvement by these special staff sections will greatly enhance the quality and realism of the training. Medical personnel will better understand OC decontamination and the possible injuries they may encounter as a result of NLW employment. Because of the international attention that NLW receive, all personnel will benefit from media training by PA personnel. CA and PSYOP personnel can provide area-specific cultural, ethnic, and political briefs to enhance awareness of the challenges within the AO. SJA personnel can clarify complex ROE questions and assist in training on ROE-related issues. Additionally, supply and logistic staff members also require training in NLW maintenance and environmental issues unique to the equipment and munitions.

5. Unit Training

The INIWIC is primarily designed to produce instructors who will train individuals in using and employing NLW. Collective unit tactical training is the unit's responsibility; however, a NLW sample schedule based on INIWIC training is provided in Appendix D. Units should conduct collective training before deploying with NLW. Unit standard operating procedures (SOP) and tactics may require adjustment to enable the employment of NLW and new equipment for the success of specific missions. All collective training should include nonlethal options and decision making. Whenever NLW are employed during training, the instructor must also conduct lethal capability training. Troops must understand the integration of nonlethal and lethal capabilities and always be prepared to protect themselves, as required, with lethal force. It is impossible to accurately predict the response of an individual or crowd.

6. Training Disclaimer

Initial unit training should not be conducted without qualified instructors. Non-type classified and COTS nonlethal munitions cannot be fired without Army Materiel Command (AMC), Army Training Support Center (ATSC), Crane Naval Weapons Center, or Air Force Munitions Center type classifying the ammunition. The instructor should conduct unit sustainment training in accordance with (IAW) service guidelines.

7. Additional Training

a. Additional training is left to the discretion of the unit but should be encouraged in order to add credibility to the training package. Some recommended training certifications for instructors include, but are not limited to—

(1) Management of Aggressive Behavior (MOAB)
R.E.B. Training International, Inc.
PO Box 845, Stoddard, NH 03464

(2) Oleoresin Capsicum Aerosol Training (OCAT)
R.E.B. Training International, Inc.
PO Box 845, Stoddard, NH 03464

(3) Monadnock Defensive Tactics System (MDTS)
Training calendar available at the Monadnock website:
http://www.batons.com/traincal.html

b. These courses add viable credentials to the instructors as perceived by civilian and military alike. In addition, the instructor can issue certification cards for the area which an individual has been certified. Certifications in various areas offers additional protection in the event of a lawsuit or claim against the government by ensuring the individuals that have been trained within your unit are completely capable.

8. Summary

NLW and tactics training are not meant to be stand-alone training—units must anticipate the requirement for NLW in their missions and train accordingly. NLW provide commanders with a graduated response in the force continuum.

Chapter V

EMPLOYMENT

1. Background

a. A military unit deployed to an area will normally conduct operations IAW current doctrine. The environment may contain both noncombatants and combatants. Nonlethal capabilities provide the military force with tools to engage adversaries and minimize collateral damage. Fundamental to employing NLW is a thorough understanding of the force continuum. As stated before, nonlethal capabilities provide a wider range of options that augment traditional means of deadly force but do not replace them. Nonlethal capabilities allow a commander the flexibility to increase and decrease the amount of force applied to accomplish a mission. The force continuum is generally continuous and seamless, yet a careful examination reveals five broad categories:

(1) Threats. The force continuum begins with threats. These can be either implied or expressed. An implied threat may be manifested through presence by forming into riot-control formations, fixing bayonets, or donning protective masks. An expressed threat occurs when a commander makes known the consequences of defiance.

(2) Denial options. The equipment used in denial options are usually placed in highly visible areas and often marked with placards. Examples include concertina, caltrops, barbed wire, or other obstacles.

(3) Munitions that cause physical discomfort/incapacitate. Munitions that cause physical discomfort, but fall short of inflicting trauma, include flash bangs, RCA—tear gas (O-Chlorobenzyl-Malononitrile [CS]), and OC. Although the discomfort or injury may be substantially less than impact munitions, employing these munitions requires an appropriate use of force assessment. Factors such as training, discipline, prejudices, emotions, and judgment all play a part in their application.

(4) Munitions that inflict blunt trauma. Munitions that inflict blunt trauma will inflict painful or debilitating injuries. These munitions constitute the upper end of nonlethal options within the force continuum. Examples include foam batons, stingball grenades, beanbags, sponge projectiles, and rubber ball rounds.

(5) Lethal munitions. Lethal weapons are at the highest level of force continuum. Lethal options should always be regarded as part of the force continuum and not as a separate option altogether. ROE, along with judgment and experience, helps on-scene commanders decide when using lethal force is authorized and appropriate.

b. Nonlethal capabilities include using existing systems that were not designed specifically as NLW. Nonlethal capabilities can minimize fatalities, permanent injury to personnel, and undesired damage to property, facilities, and the environment. Examples include PSYOP, CA, military working dogs, and PA. However, this chapter limits discussion primarily to using weapons and components explicitly designed as part of a NLW capability set. This focus is not intended to limit the commander's option to use whatever means are available to accomplish the mission without unnecessary lethality.

2. Mission Planning Factors

Incorporating NLW into a plan does not change the fundamental principles of mission planning. Commanders should use NLW as an operational enhancement. The first concern of a commander is successfully completing the mission. ROE are a critical consideration in the commander's planning and guide him in determining the appropriate use of nonlethal and lethal force. The availability of NLW does not relieve a force from its obligation to defend itself. Therefore, it must be equipped and trained with both lethal and nonlethal means.

3. NLW Tasks

a. NLW tasks were derived from the combatant commanders and service mission needs statements and are codified in the JMAA for Nonlethal Weapons. These tasks are grouped under three core capabilities: counterpersonnel, countermateriel, and countercapability.

(1) Counterpersonnel:

(a) Control crowds.

(b) Incapacitate personnel.

(c) Deny an area to personnel.

(d) Clear facilities/structures of personnel.

(2) Countermateriel:

(a) Deny an area to vehicles, vessels, and aircraft.

(b) Disable/neutralize vehicles, vessels, aircraft, and equipment.

(3) Countercapability:

(a) Disable or neutralize facilities and systems.

(b) Deny use of WMD.

b. Although not all encompassing, the following list of employment considerations is designed to assist commanders and staffs in formulating plans for use of NLW:

(1) General considerations:

(a) Understand and define ROE.

(b) Commander develops identification of friend or foe (IFF) procedures.

(c) Distribute NLW munitions to troops in advance.

(d) NLW options require lethal overwatch fire.

(e) Ensure area is well lighted to include perimeter.

(f) Employ night vision devices outside the perimeter to identify personnel.

(g) Utilize appropriate media to inform civilians to stay away from denied areas.

(h) Enforce positive access control into perimeter (personnel/vehicles).

(i) Ensure posted signs are easily understood (symbols/pictures). If interpreter support available, post warning signs in native language.

(j) Use entanglements such as caltrops, tanglefoot, barbed wire, and concertina/razor tape as these disrupt/slow intrusion attempts.

(k) Employ sensors, if available.

(l) Employ aerial or naval pickets.

(2) Counterpersonnel considerations:

(a) Employing kinetic rounds at less than 15 feet/5 meters may result in fatal outcomes.

(b) Target area at 15 to 40 feet should be center mass for blunt munitions.

(c) Head shots may be lethal.

(d) Kinetic rounds will not be skip fired.

(e) Kinetic rounds are effective against selected individual targets and can be used to disperse individuals.

(f) Considering differing rates of fires; the M203 and shotgun are complementary in a nonlethal role, depending upon the munitions used and their effective ranges.

(g) Employ riot batons only after proper training.

(h) Use RCA to disrupt/disperse crowds.

(i) Use RCA markers (RCA with a nontoxic, water-soluble marking dye) to mark agitators for future identification and apprehension.

(j) RCA greatly reduce visibility of control force.

(k) Closely control the employment of RCA.

(l) Disperse RCA to face and eyes for full effect.

(m) Employ optics with designated marksmen to identify mob leaders.

(n) Searchlights dazzle/disorient individuals (night only).

(o) Use bullhorn for communicating with crowd and to control formations.

(p) Employ recovery/apprehension/snatch teams to apprehend agitators.

(q) Use flex cuffs for agitators.

(r) Prepare disposition plan for cleared personnel.

(s) Isolate area to prevent reinforcements.

(t) Consider future friendly use of cleared facility.

(u) Prepare security plan for cleared facility.

(v) Channel passive traffic away from formations.

(w) Maintain rapid reaction force/quick reaction force (RRF/QRF).

(3) Countermateriel/countercapability considerations:

(a) Use marking agents to mark a vessel for later interception.

(b) Consider materiel handling equipment required to evacuate vessel/airframe.

c. Selected NLW tasks are further addressed in Tables V-1 through V-7. The tables are not intended to be all-inclusive but merely a stimulus to planning. The equipment listed is taken from the currently available systems designed for nonlethal purposes. The user should not limit his nonlethal options to this specific equipment and should be aware that there are many existing and emerging systems and potential uses of current technology and common objects to achieve the desired mission end state.

Table V-1. Crowd Control

Tasks	*NLW Planning Factors*	*Equipment*	*Employment Considerations*	*Notes*
Crowd control	Administration/ Personnel • SJA / claims planning • Chaplain affairs • Casualty notification • Detainee disposition plan • PAO/media planning • Information dissemination	Support Equipment • Riot face shield • Full length riot shield • Expandable riot baton • Optics • Portable bullhorn • Vest	• Distribute to troops in advance • Employ riot baton only after proper training • Employ optics with designated marksman (DM) to identify (ID) mob leaders • Use bullhorn for communicating with crowd and to control formations	• Strictly define ROE • NLW options require lethal cover fire
Crowd control	Intelligence • Intelligence preparation of the battlespace (IPB) • Background on country and culture • Route/map recon • Sensor emplacement for advance notification • Information on Mob leaders • Patrols and listening post (LP)/observation post (OP) • Determine reason for the crowd	Kinetic Rounds • 12-gauge (ga) bean bag round • 12-gauge rubber bullet • 12-gauge launching cartridge • 40-mm rubber baton • 40-mm wooden baton **NOTE:** USMC no longer uses wooden baton. • 40-mm stinger grenade • Stun grenade • Flash bang	• Do not employ kinetic rounds at less than 15 feet due to possible fatal outcome • Target area at 15 to 40 feet should be center mass • Head shots are not acceptable • Kinetic rounds will not be skip fired • Kinetic rounds effective against selected targets and to disperse individuals • M203 and shotgun are nonlethal complementary; M203 has low rate of fire; area target. Shotgun has high rate of fire; point target	

Table V-1. Crowd Control

Tasks	*NLW Planning Factors*	*Equipment*	*Employment Considerations*	*Notes*
Crowd control	Operations • Tactical Integration of NLW and lethal fires • Fire Support Element (FSE) plans to include RCA • Allocation of NLW munitions • Integration of NLW fires and maneuver • Engineer barrier planning • Understand mission (deny access/disperse/ monitor)	RCA • OC Dispenser • Team OC dispenser • High volume dispenser	• Disrupt/disperse crowds • Disperse RCA to face and eyes for full effect • Employ snatch teams to apprehend agitators	
Crowd control	Logistics • Distribution of NLW supplies • Security of rear area facilities • Resupply of NLW munitions • Procurement of medical supplies • Hospital planning • Medic planning for tactical units	Riot Control • Xenon Searchlight • Flex cuffs • RCA markers	• Searchlights dazzle/disorient individuals (night only) • Flex cuffs used for agitators • RCA markers are used to mark agitators for future ID	

Table V-2. Incapacitate Personnel

Task	*NLW Planning Factors*	*Equipment*	*Employment*	*Notes*
Incapacitate personnel	Administration/Personnel • SJA/claims planning • Media plan	Support Equipment • Riot face shield • Full length riot shield • Expandable riot baton • Rifleman's combat optic • Vest	• Distribute to troops in advance • Do not employ riot baton to the head • Employ optic with DM to ID mob leaders and cover with lethal fire	• Commander needs to define incapacitation mission • Take a balanced approach to public dissemination of NLW so counter measures cannot be rapidly developed • Cover NLW options by lethal force
Incapacitate personnel	Intelligence • ID types of insurgents and demonstrators • ID causes and factions • ID mob ringleaders	Kinetic Rounds • 12-gauge bean bag round • 12-gauge rubber bullet • 12-gauge launching cartridge • 40-mm rubber baton • 40-mm wooden baton **NOTE:** USMC no longer uses wooden baton. • 40-mm stinger grenade • Stun grenade • Flash bang • M203 and shotguns are nonlethal complementary. M203 has a low rate of fire, area target. Shotgun has a high rate of fire, point target	• Do not employ kinetic rounds at less than 15 feet due to possible fatal outcome • Target area at 15-40 feet should be center mass • Head shots are not acceptable • Kinetic rounds will not be skip fired • Kinetic rounds effective against selected targets and to disperse individuals • M203 and shotguns are nonlethal complementary. M203 has a low rate of fire, area target. Shotgun has a high rate of fire, point target	

Table V-2. Incapacitate Personnel

Task	*NLW Planning Factors*	*Equipment*	*Employment*	*Notes*
Incapacitate personnel	Operations • Prepare and confirm ROE for NLW • Integration of NLW and deadly force • Allocate NLW and designate unit use • Integrate NLW fires and maneuver	RCA • OC dispenser • Team OC dispenser • High-volume • OC dispenser	• Disrupt/disperse crowds • Disperse RCA to face and eyes for full effect	
Incapacitate personnel	Logistics • Distribute NLW munitions • ID special storage requirements • Resupply NLW munitions	Riot Control • Xenon • searchlight • Flex cuffs • Chemical markers	• Searchlights dazzle/disorient individuals (night only) • Use flex cuffs for agitators • Use RCA markers to mark agitators for incapacitation	
Incapacitate personnel	Civil Affairs • Collect civilian intelligence • Care for injured personnel • Detain civilian personnel			

Table V-3. Deny Area to Personnel

Task	*NLW Planning Factors*	*Equipment*	*Employment*	*Notes*
Deny area to personnel	Administration/ Personnel • SJA/claims planning • Media plan	Support Equipment • Area lights • Area sensors • Night vision • Access control • Warning signs written in host nation language	• Ensure area is well lighted to include perimeter • Employ night vision devices outside the perimeter • Positively ID accessed personnel • Employ warning signs in native language	• Utilize appropriate media to warn civilians away from denied area

Table V-3. Deny Area to Personnel

Task	*NLW Planning Factors*	*Equipment*	*Employment*	*Notes*
Deny area to personnel	Intelligence • Terrain and map study • Threat ID • Sensor emplacement around area • IPB			
Deny area to personnel	Operations • Prepare and confirm ROE • Integrate NLW and deadly force • Allocate NLW and designate unit use • Determine scope of mission • Determine access procedures • Use of area (friendly or total denial)			
Deny area to personnel	Logistics • Distribute NLW munitions • ID special storage requirements • Resupply NLW munitions • Security of rear area facilities.			
Deny area to personnel	Civil Affairs • Collect civilian intelligence	Entanglements • Caltrops • Tanglefoot • Barbed wire • Foams • Concertina/ razor tape • Stinger spike system	• Disrupt/slow intrusion attempts • Maintain suitable reaction force • Channel passive traffic away • Use foam to cover caltrops	

Table V-4. Deny Area to Vehicles

Task	*NLW Planning Factors*	*Equipment*	*Employment*	*Notes*
Deny area to vehicles	Administration/ Personnel • SJA/claims planning • Media plan	Support Equipment • Area lights • Area sensors • Night vision • Access control • Warning signs written in host nation language	• Ensure area is well lighted to include perimeter • Employ night vision devices outside the perimeter • Positively ID accessed vehicles	• Commander needs to define denial mission • Utilize appropriate media to warn civilians away from denied area
Deny area to vehicles	Intelligence • Terrain and map study • Threat ID • Sensor emplacement around area • IPB • Traffic analysis			
Deny area to vehicles	Operations • Prepare and confirm ROE • Allocate NLW and designate unit use • Determine scope of mission • Determine access procedures • Determine MP/ security forces requirements • Use of area (friendly or total denial)			• Commanders must consider flying operations when employing RCA. Area(s) should be reusable in short order.
Deny area to vehicles	Logistics • Distribute NLW supplies • Be prepared to evacuate stalled vehicles			
Deny area to vehicles	Civil Affairs • Collection of civilian intelligence	Entanglements • Caltrops • Tanglefoot • Barriers • Organic equipment • Barbed wire • Foams	• Disrupt/slow intrusion attempts • Maintain suitable reaction force • Channel passive traffic away • Use foam to cover caltrops	• Commanders must consider flying operations when employing NLW. NLW should not hinder flying operations.

Table V-5. Deny/Neutralize Vehicles, Aircraft, Vessels, and Facilities

Task	***NLW Planning Factors***	***Equipment***	***Employment***	***Notes***
Deny/ neutralize vehicles, aircraft, vessels, facilities	Administration/Personnel • SJA/claims planning • Media plan	Support Equipment • Sensors to track approaching vehicles, aircraft, and vessels • Night vision equipment • Materiel handling equipment • Portable bullhorn	• Employ ground and water sensors, if required • Employ aerial or naval pickets • Use materiel handling equipment to evacuate vessel	• Commander needs to define neutralization mission • Element of surprise remains a critical factor in mission accomplishment • Commander needs to develop IFF procedures
Deny/ neutralize vehicles, aircraft, vessels, facilities	Intelligence • Sensor emplacement if required • Threat ID • IPB • Analyze target area			
Deny/neutralize vehicles, aircraft, vessels, facilities	Operations • Prepare and confirm ROE • Integrate NLW and deadly force • Allocate NLW and designate unit use • Determine scope of mission • Determine access procedures • Determine boarding procedures (vessels and aircraft) • Determine apprehension procedures • Determine decontamination (DECON) procedures	RCA • Marking agent	• Use marking agents to mark a vessel for later interception	

Table V-5. Deny/Neutralize Vehicles, Aircraft, Vessels, and Facilities

Task	*NLW Planning Factors*	*Equipment*	*Employment*	*Notes*
Deny / neutralize vehicles, aircraft, vessels, facilities	Logistics • Distribute NLW munitions • ID special storage requirements • Resupply NLW munitions • Determine vessel/aircraft disposition	Riot Control • Xenon searchlight	• Use RCA to incapacitate crew members • Searchlights can disorient personnel or illuminate target	• Commanders must consider flying operations when employing RCA • Area(s) should be reusable in short order.
Deny / neutralize vehicles, aircraft, vessels, facilities	Civil Affairs • Collect civilian intelligence • Handle displaced civilian personnel	Entanglements • Caltrops • Foams • Speed bumps • Nets • PVAB	• Use caltrops to disable vehicle • Use combination of foam and caltrops to preserve the element of surprise • Channel passive traffic away • Use nets to foul props	
Deny/neutralize vehicles, aircraft, vessels, facilities	Medical • Treating injured civilian personnel			

Table V-6. Seize Personnel

Task	*NLW Planning Factors*	*Equipment*	*Employment*	*Notes*
Seize personnel	Administration/ Personnel • SJA/claims planning • PAO media plan	Support Equipment • Optics	• Employ optics with DM to ID personnel to be seized • Cover personnel employing nonlethal force	• Commander needs to define seizure mission and strictly enforce ROE • Element of surprise is critical for mission accomplishment • Speed is critical • Seizing personnel can be surgical with special operations personnel or conducted within a crowd to seize an agitator
Seize personnel	Operations • Prepare and confirm ROE • Integrate NLW and deadly force • Allocate NLW and designate unit use • Integrate NLW fires and maneuver • Breaching and assault planning • Determine holding area/processing center	RCA • OC dispenser • CS	• Use RCA to incapacitate personnel • Spray into eyes and face for full effect	
Seize personnel	Logistics • Distribute NLW munitions • ID special storage requirements • Resupply NLW munitions • Plan transportation	Riot Control • Xenon searchlight • Flex cuffs • RCA markers	• Searchlights dazzle/disorient individuals • Use flex cuffs for apprehension • Use RCA markers to mark agitators for seizure	

Table V-6. Seize Personnel

Task	*NLW Planning Factors*	*Equipment*	*Employment*	*Notes*
Seize personnel	Civil Affairs • Collect civilian intelligence • Care for injured personnel • Detain civilian personnel			
Seize personnel	Medical • Medical evacuation (MEDEVAC) planning			

Table V-7. Clear Facilities of Personnel

Tasks	*NLW Planning Factors*	*Equipment*	*Employment Considerations*	*Notes*
Clear facilities of personnel	Administration/ Personnel • SJA/claims planning • Chaplain affairs • Casualty notification • PAO media plan	Support Equipment • Riot face shield • Full-length riot shield • Expandable riot baton • Optics • Portable bullhorn	• Distribute to troops in advance • Do not employ riot baton to the head • Employ optic with DM to ID occupant leaders and provide lethal cover fire • Use bullhorn used for communicating with occupants	• Surprise is a critical element for mission accomplishment • Commander must strictly define ROE • Due to close quarter tactics, the force must be instantly ready to utilize deadly force if required

Table V-7. Clear Facilities of Personnel

Tasks	***NLW Planning Factors***	***Equipment***	***Employment Considerations***	***Notes***
Clear facilities of personnel	Intelligence • IPB • Background on facility, access, power plant • Route/map recon • Background info on building occupants • Determine any outside support for building occupants	Kinetic Rounds • 12-gauge bean bag round • 12-gauge rubber bullet • 12-gauge launching cartridge • 40-mm rubber baton • 40-mm wooden baton **NOTE:** USMC no longer uses wooden baton. • 40-mm stinger Stun grenade • Flash bang	• Do not employ kinetic rounds at less than 15 feet due to possible fatal outcome • Targeted personnel at 15 to 40 feet should be center mass • Head shots are not acceptable • Kinetic rounds will not be skip fired • Kinetic rounds effective against selected targets and to disperse individuals • Stun grenade effects can be amplified in buildings • Visibility will be reduced once kinetic rounds are employed • M203 and shotguns are nonlethal complementary. M203 has a low rate of fire, area target. Shotgun has a high rate of fire, point target	• Using RCA greatly reduces visibility • Use of RCA is restricted to certain circumstances; verify that proper authority has been given before use

Table V-7. Clear Facilities of Personnel

Tasks	*NLW Planning Factors*	*Equipment*	*Employment Considerations*	*Notes*
Clear facilities of personnel	Operations • Tactically integrate NLW and lethal fires • FSE plans to include RCA • Allocate NLW munitions • Integrate NLW fires and maneuver • Determine assault and breaching plan • Plan for isolation of facility	RCA • OC dispenser • Team OC dispenser • High-volume dispenser	• Disrupt/disperse crowds • Apply RCA to face and eyes for full effect	• Isolate area to prevent reinforcements • Prepare disposition plan for cleared personnel • Consider use of facility • Prepare security plan for cleared facility
Clear facilities of personnel	Logistics • Distribute NLW supplies • Resupply NLW munitions • Plan for transportation of displaced civilians • Plan facility maintenance or repair	Riot Control • Xenon searchlight • Flex cuffs • RCA markers • Flash bangs	• Searchlights dazzle/disorient individuals • Use flex cuffs for agitators • Use RCA markers to mark agitators for future ID • Flash bangs to dazzle/disorient individuals	
Clear facilities of personnel	Civil Affairs • Collect civilian intelligence • Care for injured civilians • Coordinate plans for controlling civilians			
Clear facilities of personnel	Medical • Hospital planning • Medic planning for tactical units • Procurement of medical supplies			

4. Staff Planning/Responsibilities

a. Administration. A staff composed of the G-1/S-1, PAO, SJA, and others as needed, is responsible for advising the commander on NLW employment and ROE. Additionally, administrative personnel are responsible for processing all civilian detainees and claims that arise from NLW employment. As a minimum, the staff must consider the following:

(1) SJA/claims planning.

(2) Chaplain.

(3) Casualty notification.

(4) Detainee disposition plan.

(5) PAO/media planning.

(6) Information dissemination.

b. Intelligence. The G-2/S-2 office is responsible for conducting the IPB process and providing information on the adversary's culture, political stability, and level of dedication to the cause, motivation, leadership, capabilities, equipment, and tactics. The G-2/S-2 should locate native language speakers to assist the interrogator-translator team.

(1) Background on country and culture.

(2) Route/map reconnaissance.

(3) Place sensors for advance notification.

(4) Information on leaders.

(5) LP/OP.

(6) Determine reason for the crowd.

(7) Identify threat.

(8) Analyze traffic.

c. Operations. The G-3/S-3 office is responsible for unit training, nonlethal certification of units, mission analysis, maneuver plans, and CA planning when required.

(1) Tactically integrate NLW and lethal fires.

(2) Fire support plans to include using RCA.

(3) Allocate NLW munitions.

(4) Plan engineer barrier.

d. Logistics. The G-4/S-4 office is responsible for logistics considerations, ship-to-shore movement of weapons and ammunition, storage and maintenance of equipment, resupply, transportation ashore, and other support requirements inherent in using NLW.

(1) Distribute NLW supplies.

(2) Security of rear area facilities.

(3) Resupply NLW munitions.

(4) Special maintenance requirements for NLW equipment and munitions.

e. Communications Officer. The G-6/S-6 staff is responsible for installing, operating, and maintaining communications from units confronting belligerents to the tactical operation center (TOC) or combat operations center (COC). The staff must plan for the effects NLW will have on communications to include proper frequencies.

f. PAO. The PAO provides liaison with outside media and participates in the staff planning process. PAO is also responsible for ensuring the NLW media plan and news releases consider the perceptions of the native media and belligerents or adversaries. PA personnel must receive training on NLW technologies, employment and capabilities, acute and long-term effects, impact on the environment, and legal approval procedures (at a minimum) so they can respond to media and public inquiries.

g. Medical. Medical personnel must be aware of the effects of NLW. Medical personnel should also be familiar with and able to provide decontamination of RCA as well as administer antidotes in cases of adverse reactions to riot control and other agents.

(1) MEDEVAC planning.

(2) Procuring medical supplies.

(3) Hospital planning.

(4) Medical planning for tactical units.

h. Chaplain. The commander should consider chaplain support for counseling and/or religious services for military and civilian personnel. Additionally, chaplains should provide information regarding the religious and cultural background of the local populace.

i. CA. The CA office is responsible for planning and coordinating all aspects of CA activities and providing assistance in gathering intelligence from civilian sources.

(1) Handle displaced civilian personnel.

(2) Coordinate plans for control of civilians.

j. PSYOP. Tactical PSYOP forces serve as an ideal tool for the commander to resolve issues of armed resistance and local interference with military operations while encouraging dialogue and cooperation with noncombatants. This force provides dissemination of PSYOP products (loudspeaker messages, leaflets, posters, etc.) and conducts face-to-face communications with target audience. PSYOP forces provide the supported commanders with nonlethal leverage by influencing foreign populations and expressing information subjectively to influence attitudes and behavior, thereby obtaining compliance or noninterference. They provide public information to foreign populations to support humanitarian activities, ease suffering, and restore or maintain civil order. They serve as the supported commander's voice to foreign populations to convey intent, countering enemy propaganda, misinformation, and opposing information to correctly portray friendly intent and actions. Finally, PSYOP forces deny others the ability to polarize public opinion and political will against the United States and its allies within an AO.

5. Tactics

The commander should include the following considerations in preparing for unit tactical operations involving NLW:

a. Avoid reconfiguring small-unit organization, except for special configurations such as—

(1) Special teams.

(2) Attachments.

(3) Crowd control formations.

(4) Security organizations.

b. Standardize postures in unit tactical SOP. Take the following steps to add a higher volume of specialty impact munitions and equipment within squads, platoons, or units:

(1) Avoid arming troops involved in apprehending designated personnel with long weapons (i.e., shotguns, rifles) because they can be easily grabbed/pulled away by individuals.

(2) Add weapons systems (shotguns) that provide a greater nonlethal capability. When loaded with nonlethal munitions, the longer reloading time of the M203 grenade launcher is protected by coverage from the high rate of fire from shotgun.

(3) Add nonlethal munitions to existing organic weapons systems (M203 grenade launcher).

(4) Avoid arming troops in the front lines of crowd control formations with long weapons.

(5) Focus riot-control capabilities in specialized reaction forces.

(6) Establish base line NLW postures.

(7) Establish command relations for special units.

(8) Maximize distance and barriers.

(9) Create NLW range cards for static positions.

(10) Provide the force with available OC/RCA consistent with ROE.

(11) Always mutually support forces utilizing NLW with lethal means.

(12) Consider environmental conditions (heat/cold/wind) and their effect on the performance of nonlethal munitions. Proper storage, periodic inspections, and rotation of NLW ammunition are critical to maintaining the effectiveness and viability of nonlethal munitions.

c. Unit Weapons, Unit Configurations, and Ammunition Considerations.

(1) Individual. Designate individuals as nonlethal shooters. Nonlethal shooters should also carry lethal munitions for use as required. Lethal rounds must be marked and carried in a manner to prevent confusing nonlethal rounds with lethal rounds.

(2) Squad. Squad leaders should dictate who carries nonlethal grenades and maintain their weapons loaded with lethal ammunition. Ideally, the squad should not change its task organization to accommodate nonlethal equipment, but should designate nonlethal shooters.

(3) Patrols. Commanders should not plan a nonlethal patrol, but should plan a combat/security patrol with a nonlethal capability when mission dictates.

(4) Static Positions. Individuals in static positions should have their weapons loaded with lethal munitions (mission, enemy, terrain and weather, troops and support available–time available and civil considerations dependent) (METT-T/METT-TC). Additionally, they should carry nonlethal munitions. Both shotgun and M203 work well from static positions and should complement each other in a nonlethal mode.

(5) Recovery Teams. Recovery teams have the primary responsibility of retrieving and controlling designated personnel from the crowd; the commander should establish a recovery team before deploying. The force should train team members in unarmed self-defense, open-hand control, and hand cuffing/flex cuffing. It should equip recovery teams with personal protection gear, 9-mm weapon, and flex cuffs. Due to the physical nature of recovery teams, it is important not to arm team members with long rifles. If security for the recovery team is an issue, the commander should attach a security element.

(6) Crowd Control Formations. The unit should establish crowd control formations within a minimum response time. Because of the physical nature of crowd control, individuals in these formations, if possible, should not carry long rifles. These formations should have inherent nonlethal capability. The unit must provide lethal coverage for the entire formation.

d. Individual Elements of a Formation. These elements are: base element, support element, and command element. The formation needs to provide versatility. Restrictions must never be set on numbers, equipment, or changes to the standard formations.

(1) Base Element. Role of the base element (front line):

(a) Base element serves as the front line or the base/foundation of the formation (can be divided into two or three sections or ranks).

(b) Front line consists of the shield holders.

(c) Angle the bottom of the shield to channel flammable liquids away from the body.

(d) Distance or interval between the shields differs based upon the range to the crowd. The closer the crowd, the tighter the interval between the shields.

(e) Provide frontal security.

(f) Hold a solid line.

(g) Provide intimidation factor.

(h) Fire team leaders are in the line holding shields.

(i) Secondary (tertiary) line(s) can consist of the baton holders and/or nonlethal gunners.

(j) Baton holders will—

- Strike with batons.
- Spray OC or disperse smoke.
- Serve as spotters for the NLW gunners.
- Control and detain subjects pulled from the crowd.

(k) Nonlethal gunners will—

- Consist of team leader with M203/M16 (provides nonlethal cover with M203 40-mm munitions and lethal cover with M16).
- Position themselves between shields working with baton holders.
- Serve as immediate control for the team.
- Maintain continuous muzzle awareness of the weapon.
- Provide immediate support to the shields.

(2) Support Element. Role and employment of the support element:

(a) Provide the means to employ lethal and nonlethal capabilities.

(b) Can consist of two ranks: M16/M203 (2nd rank) Mossberg 12-gauge shotguns (3rd rank).

(c) Provide security for search and recovery team.

(d) Consists of munitions teams and/or recovery teams.

(e) Munitions teams provide the means to employ lethal and nonlethal capabilities consisting of M16/M203 and Mossberg 12-gauge shotguns.

(f) Recovery teams provide ability to detain and search rioters downed by nonlethal munitions. Recovery teams will—

- Extract vehicles or personnel from the crowd who are in immediate danger.
- Detain and escort downed rioters to rear of the formation.
- Restrain and search detained individuals.
- Communicate with the base element.
- Spray or employ RCA as needed.
- Fill in the gaps in the front line as needed.
- Insert into a confined or congested area too large for the formation.
- Use a soft base line that capitalizes on the munitions fired to deploy forward of the shields in order to drive the crowd back, retrieve personnel/gear, and apprehend key riot personnel. Employed in pairs (1 x shield man/1 x shotgun).

NOTE: Refer to Marine Corps Security Force (MCSF) formation examples.

(3) Command Element. Role of the command element is to provide command and signal to the formation and consists of the platoon leader/flight commander/ platoon sergeant/flight sergeant, and squad leaders.

(a) Platoon Leader/Flight Commander, Platoon Sergeant/Flight Sergeant—

- Pass verbal commands.
- Maintain order within formations.
- Gauge and evaluate use of munitions and resupply considerations.

(b) Squad Leaders—

- Effectively control the squad.
- Pass information up the chain.
- Employ OC as directed.

(c) Designated Marksman (DM). During a nonlethal engagement, using DM provides confidence and safety to those facing a riot. The DM in an overwatch position scans the crowd, identifies threats and designated personnel for recovery, and fires lethal rounds if a lethal threat is presented. Additionally, DM are ideally suited for flank security and counter sniper operations.

6. Employment Objectives

a. Innovative employment methods of NLW and ingenuity take on critical significance. NLW are employed in concert with other nonlethal and lethal systems to create a synergistic effect.

b. The measures for countering many nonlethal options are usually apparent, quickly learned, and readily available. Because they are not intended to kill, nonlethal options "teach" an adversary what to avoid in the future. Small unit commanders must prepare to keep one step ahead of motivated belligerents.

c. Many nonlethal munitions have both maximum effective and minimum safe ranges. Individuals short of the minimum safe range may suffer severe injuries or death while the effects of most nonlethal munitions are greatly decreased at longer ranges. To be effective, an adversary must be engaged within the optimum range of the munition. Figure V-1 illustrates the ranges for a variety of nonlethal munitions and systems.

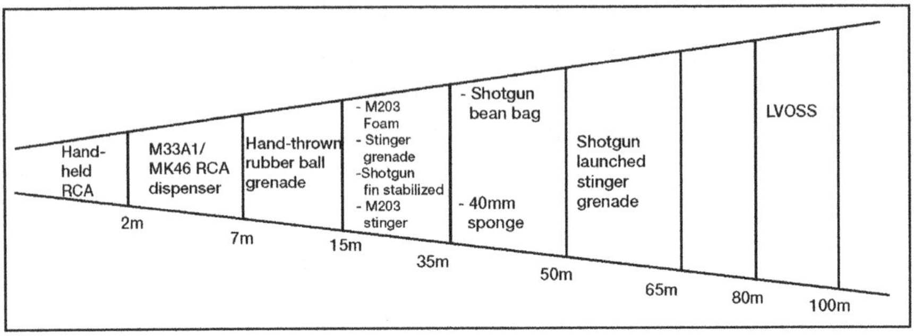

Figure V-1. Nonlethal Weapons Ranges

7. Rules of Engagement

a. U.S. forces follow directives issued by competent military authority that delineate the circumstances and limitations under which deadly force will be initiated against other forces. Intrinsic to the effective use of nonlethal options is the authorization to employ them. ROE are the primary means by which the chain of command can provide guidance to deployed forces in peacetime for handling crises and, in wartime, to help control the level of hostilities.

b. The commander should distinguish ROE from tactical control measures, threat conditions, and arming orders/weapons conditions. Although arming orders are sometimes listed on the same quick reference card as ROE for easy reference, the commander should distinguish ROE from other policies and directives developed, such as weapons confiscation rules.

c. The commander should delegate the decision to use NLW against an adversary to the lowest possible level, preferably to the platoon, flight, or squad level. However, this requires that all personnel, not just leaders, have a clear understanding of ROE and his intent.

d. Using lethal force, employed under the standing ROE, will never be denied. At no time will forces be deployed without the ability to defend themselves against a lethal threat nor will they forego normal training, arming, and equipping for combat. Nonlethal options are a complement to, not a replacement for, lethal force and seek to expand a proactive response across the range of military operations.

e. The following figures are examples of service-specific formations. These examples are not all-inclusive, and the tactical situation will always dictate actual unit employment.

FM 19-15, *Civil Disturbance*; MCWP 3-11.1, *Rifle Company;* and MCWP 3-11.2, *Marine Rifle Squad* provide extensive detail about small-unit formations, their use, and other considerations. Figures V-2 to V-7 are from I and II Marine Expeditionary Force (MEF) Special Operations Training Groups (SOTG) and represent a notional infantry platoon. Figures V-8 to V-13 represent a notional U.S. Army infantry platoon. Figures V-14 to V-21 are additional platoon formations submitted by Marine Corps Security Force Training Company.

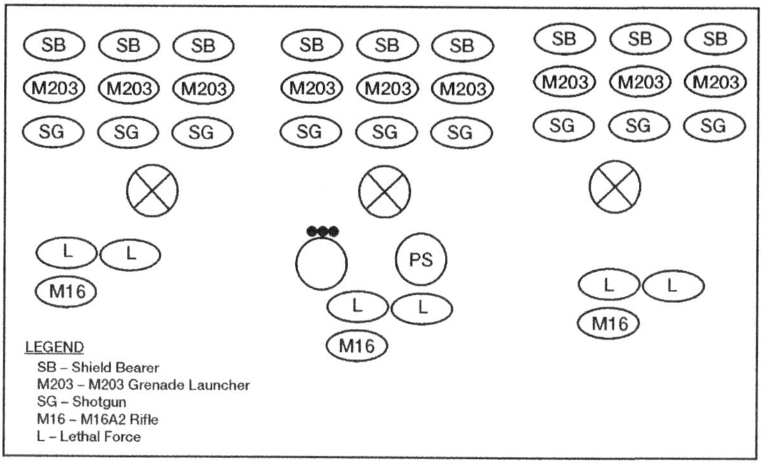

Figure V-2. Platoon in Depth

NOTE: The platoon is the smallest unit capable of independent maneuver. The platoon in depth formation shown above provides the following benefits:

- In-depth support with three ranks.
- Complementary support with the shotguns covering the slower reload time for the M203.
- Additional personnel from each squad to form the recovery teams.
- Each squad is capable of moving as an independent formation forward of the platoon line, if situation dictates.
- Other tactical formations are applicable—echelon, wedge, column.

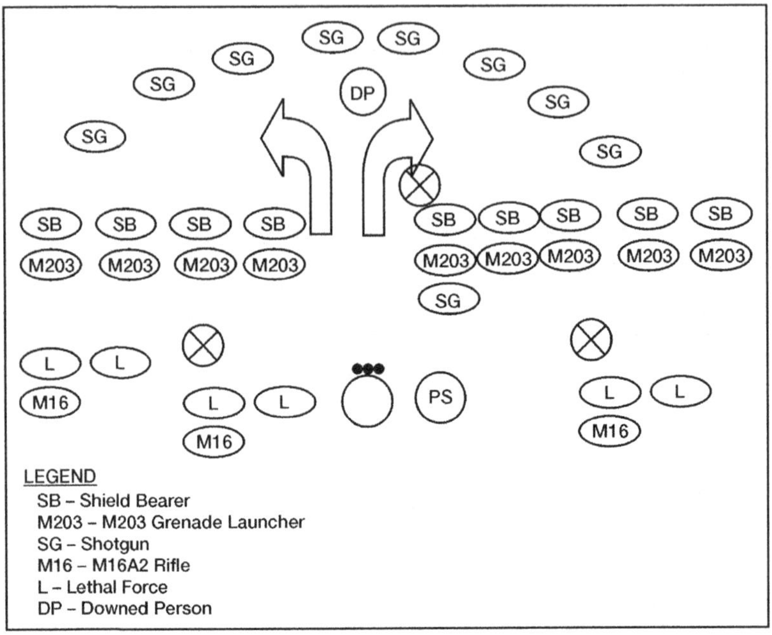

Figure V-3. Platoon Recovery Team Deployment—Security Element

---

NOTE: Security element deploys around downed person.

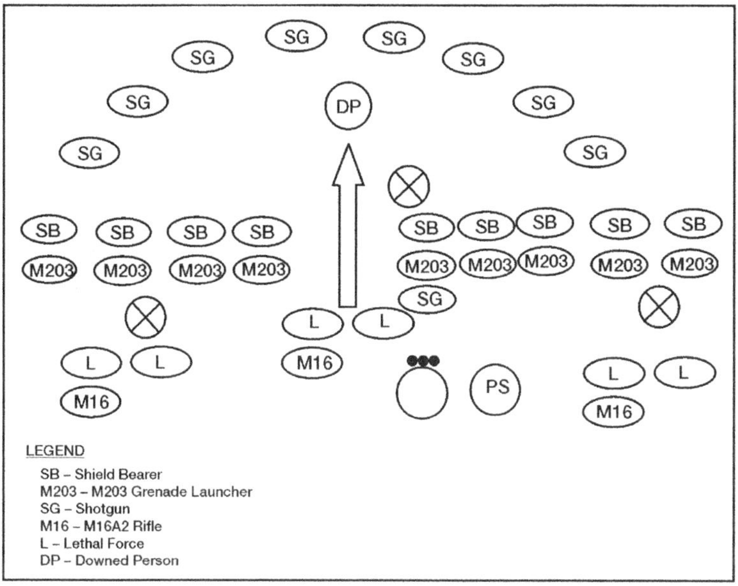

Figure V-4. Platoon Recovery Team Deployment—Recovery Team

NOTE: Recovery team deploys to detain, search, and recover downed person.

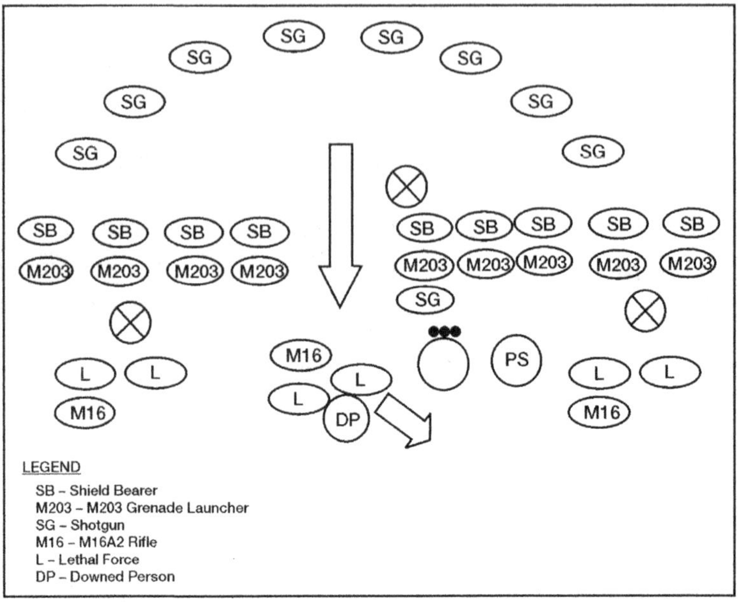

Figure V-5. Platoon Recovery Team Deployment—Recovery Team Redeploy

NOTE: Recovery team redeploys with detained person.

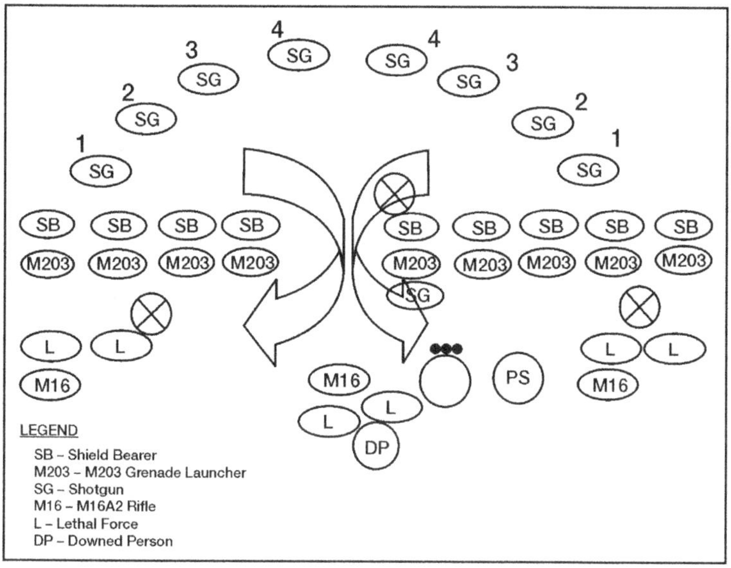

Figure V-6. Platoon Recovery Team Deployment—Security Element Redeploy

---

NOTE: Security element redeploys back to 3rd rank "by the numbers." Team leader calls "4s in," then "3s in," etc. The last person in calls "last man in" and formation closes the gap.

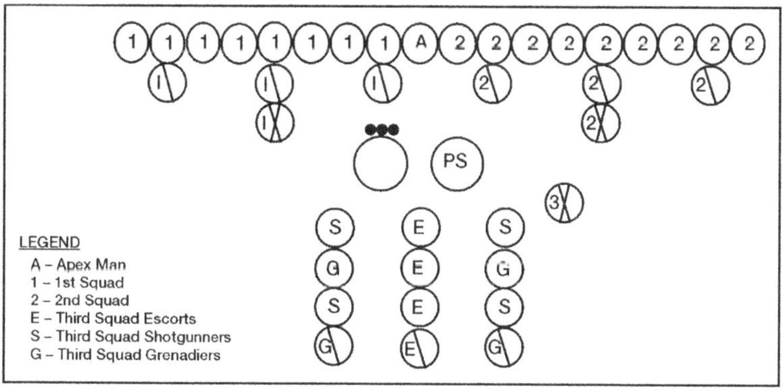

Figure V-7. Platoon On Line—Munitions Squad in Close Support to Cover Flanks

---

NOTE: Platoon in close support. NLW and lethal weapons are sited to cover the flanks of the formation, as well as support the front line.

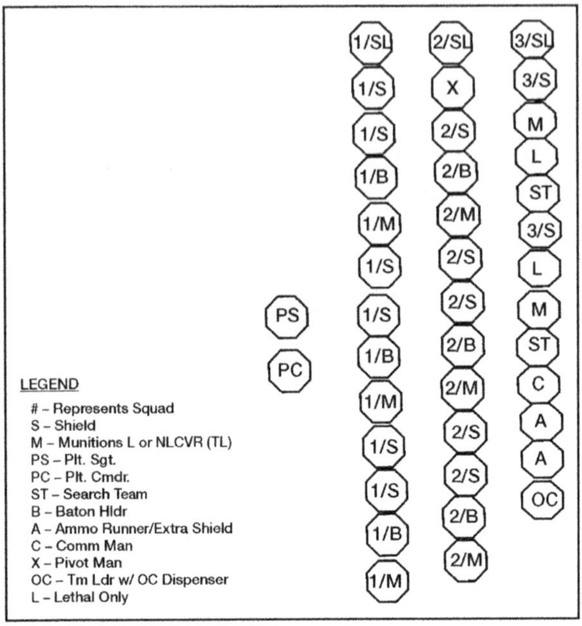

Figure V-8. Platoon Column

NOTE: Platoon in column formation. For administrative movement only.

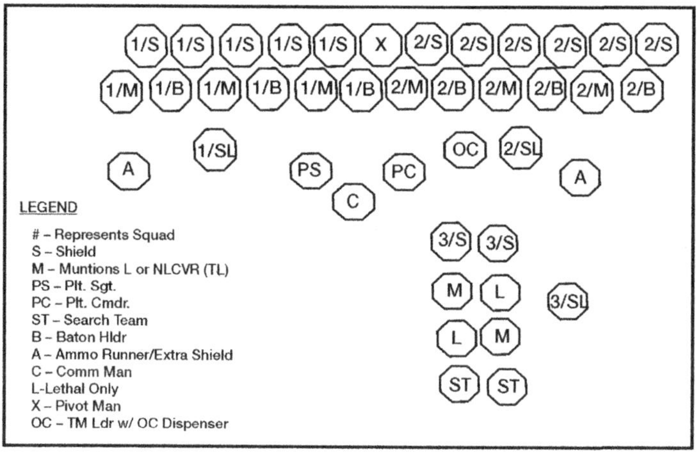

Figure V-9. Platoon On Line

NOTE: Platoon on line—basic formation for the employment of NLW.

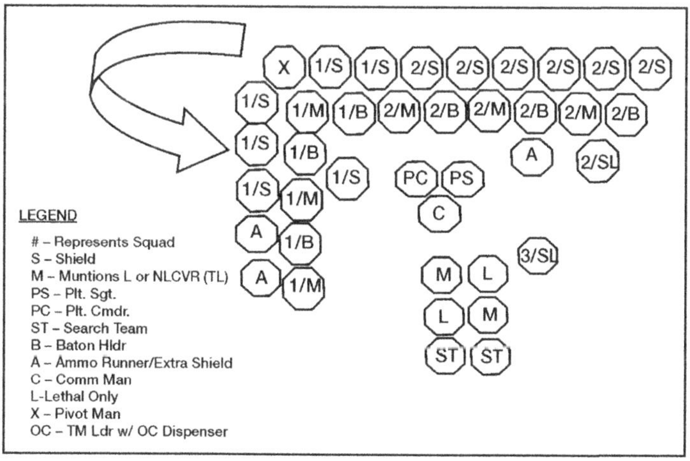

Figure V-10. Platoon Lateral Support, Left (Right)

NOTE: Platoon in lateral support to protect flanks.

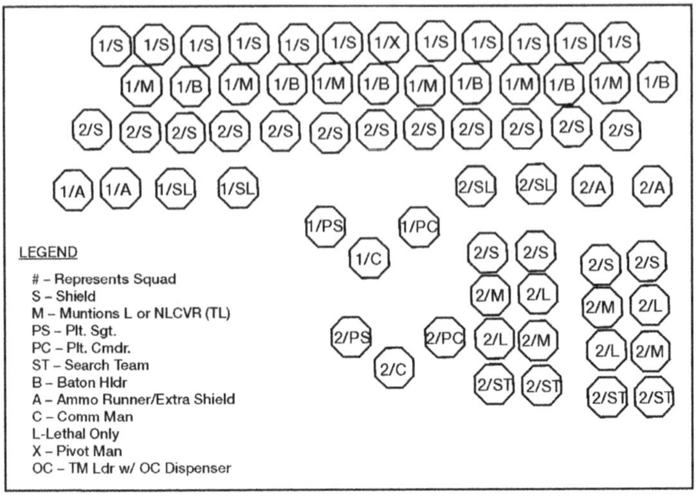

Figure V-11. Two Platoons in Close Support

***NOTE*:** Two platoons in close support—tied in for mutual advantage.

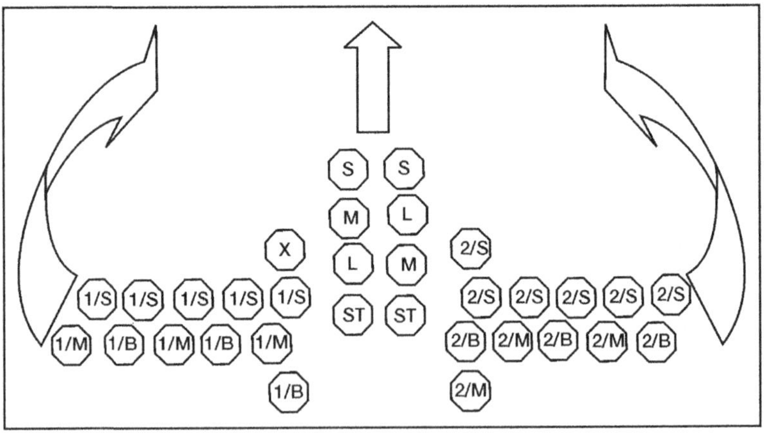

Figure V-12. Recovery Team Deployment

***NOTE*:** Recovery team deploys from formation—remainder of platoon can deploy around the recovery team from either flank.

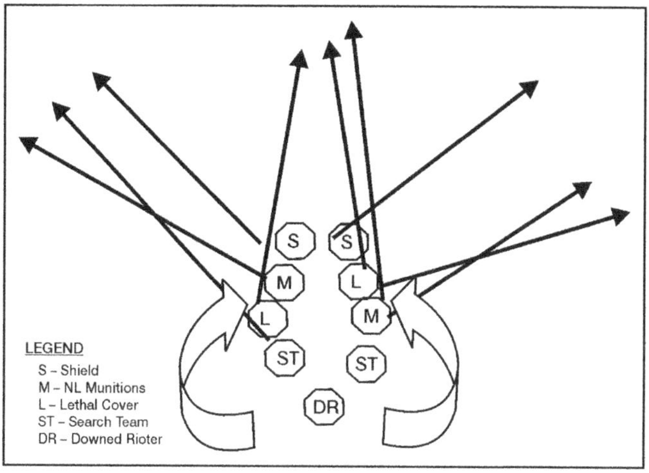

Figure V-13. Recovery Team Actions and Security Posture

NOTE: Recovery team forms security perimeter around downed rioter, conducts search and detain actions. Once actions are complete, recovery team redeploys.

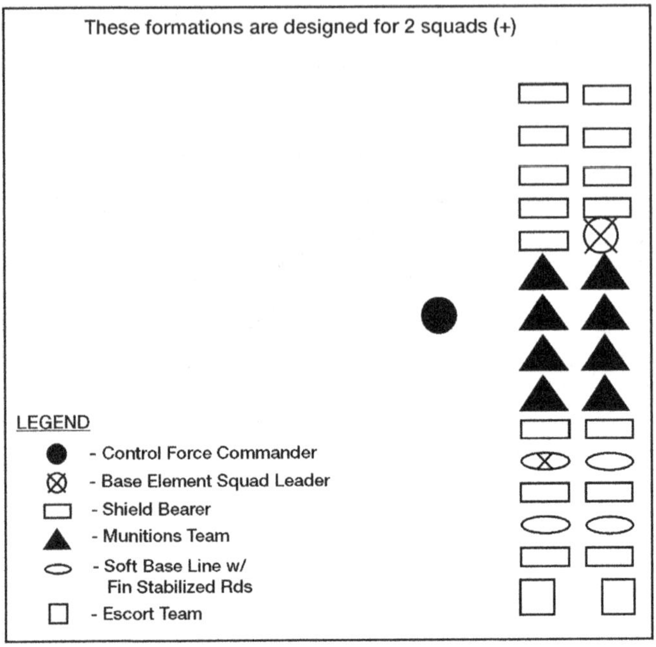

Figure V-14. Platoon in Column (USMC)

NOTE: These formations are designed for 2 squads (+). This allows the platoon to move rapidly when there is no chance of being engaged with the crowd.

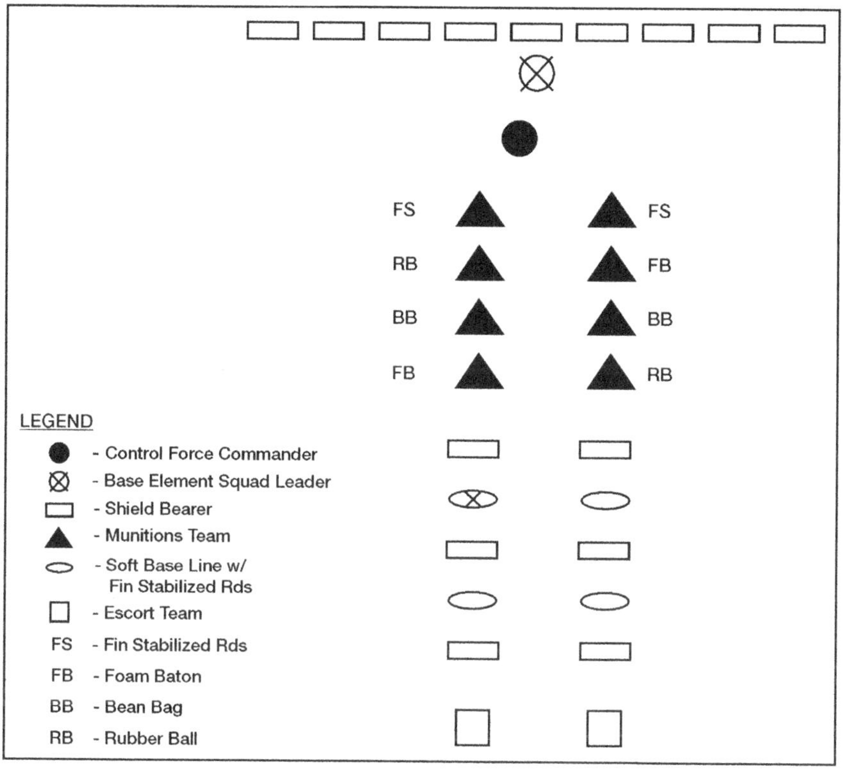

Figure V-15. Platoon in General Support (USMC)

NOTE: The control force commander moves into this formation approximately 60 to 70 meters from the crowd, as the crowd could start to engage the control force with thrown objects.

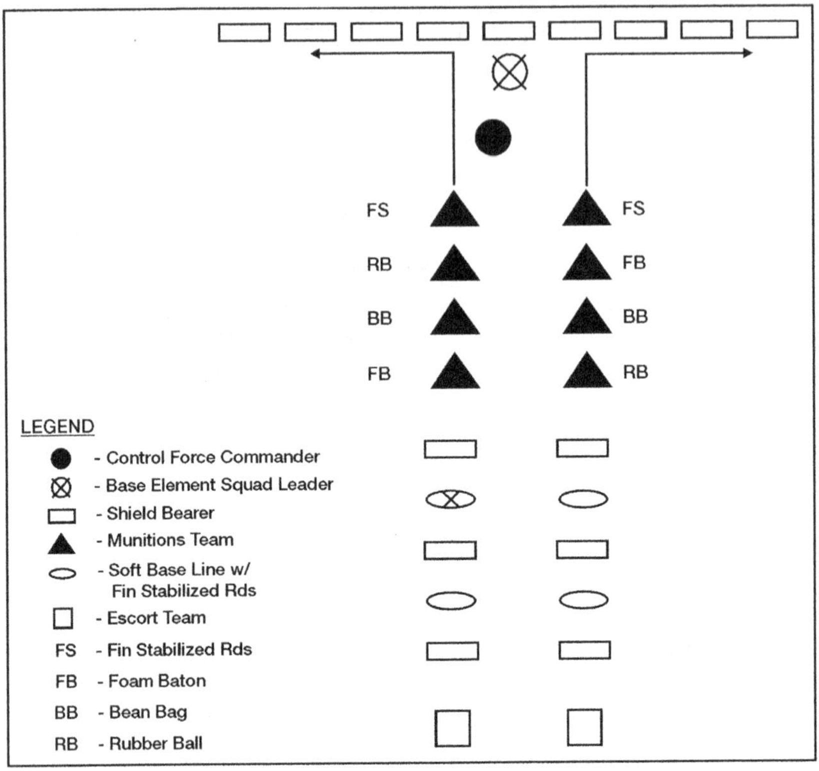

Figure V-16. Platoon in Close Support (Start) (USMC)

NOTE: The control force commander moves into this formation when he feels ready to engage the crowd with nonlethal munitions.

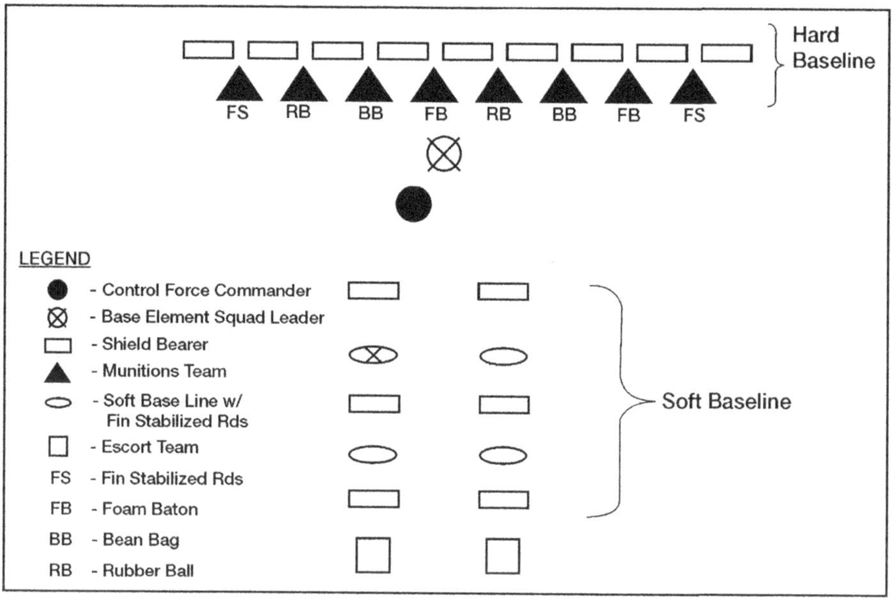

Figure V-17. Platoon in Close Support (Finish)

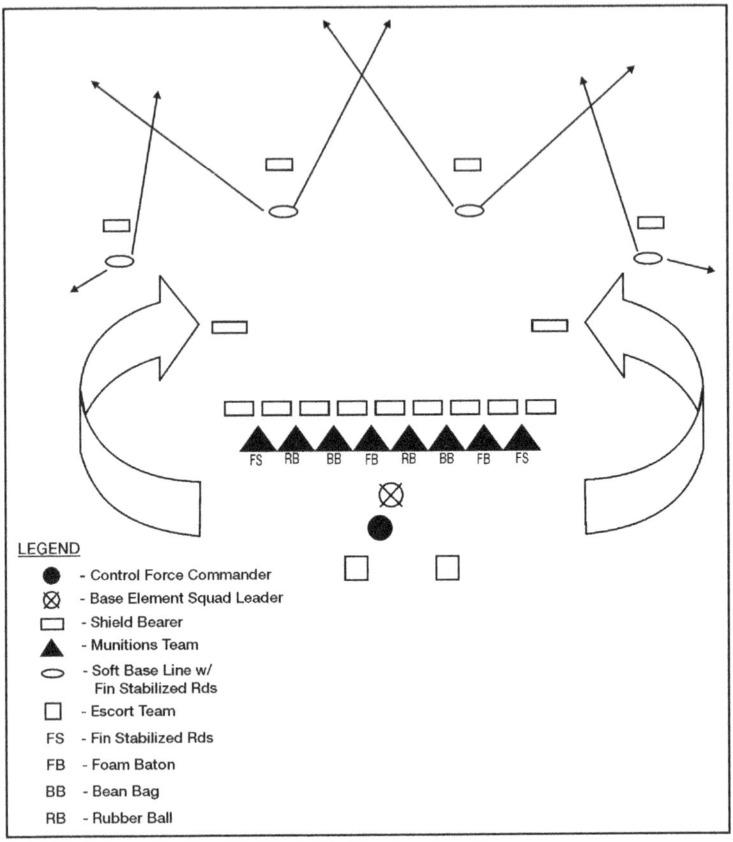

Figure V-18. Deploying Soft Baseline

NOTE: The soft baseline affords the control force commander much flexibility. The primary mission of this formation is to disperse the crowd back using speed and extreme violence of action. By deploying the soft baseline, the commander is able to maneuver the hard baseline where needed (around the barrier plan or obstacles.) The soft baseline is deployed by a predetermined signal (bullhorn siren, whistle blasts, vehicle horn) and will be retrieved using the same signal. This signal conditions the crowd to expect violence of action and causes them to disperse. The soft baseline should never be deployed into a fresh crowd; the crowd needs to be distracted by nonlethal munitions. The formula found to work is distract with stingball grenades, followed by a volley of shotguns and M203 munitions, then send the soft baseline out to disperse or move the crowd back. (Sting them, shoot them, soften them up.)

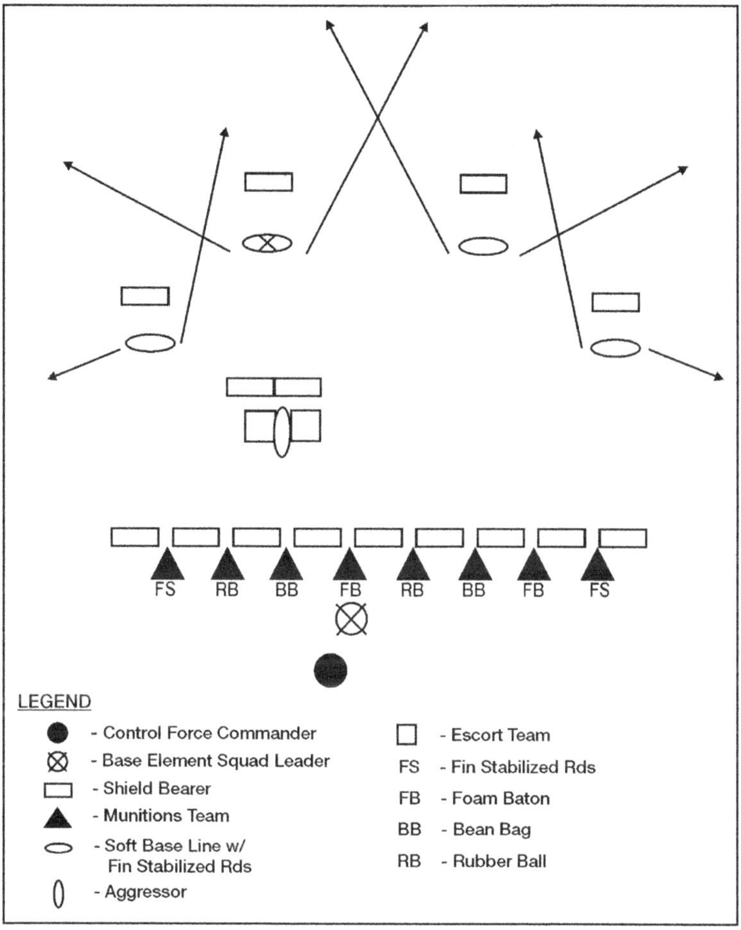

Figure V-19. Deploying Escort Team (Start)

NOTE: When the control force commander identifies the aggressor to retrieve, the command is, "Escort team left/right," then the signal. The signal is the same to deploy the soft baseline. As the soft baseline deploys, the two shields without gunners provide protection for the escort team. The shields remain with the escort team until the soft baseline is called back in, at which time the shields go back to the soft baseline.

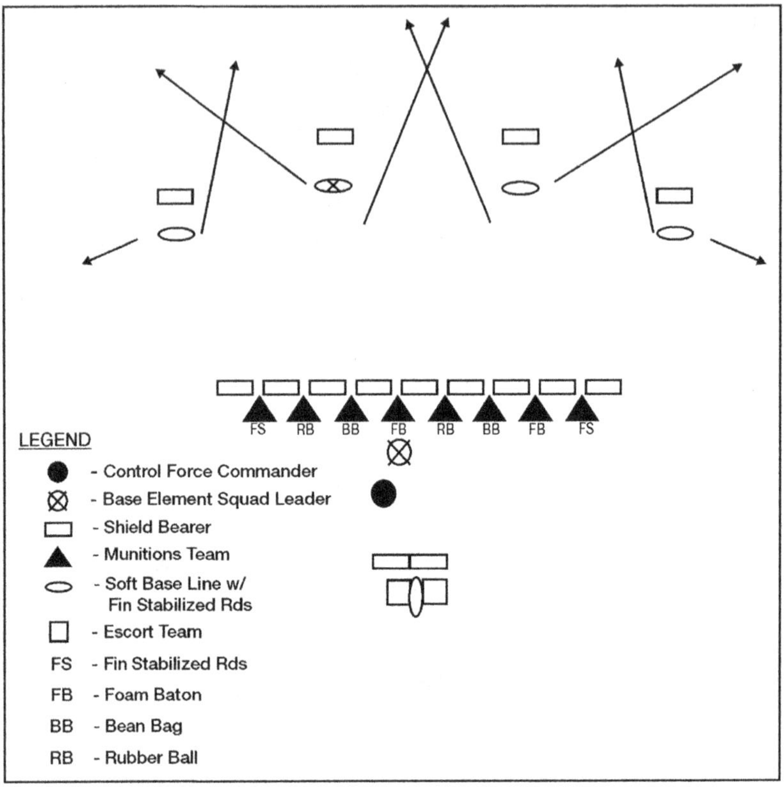

Figure V-20. Deploying Escort Team (Finish)

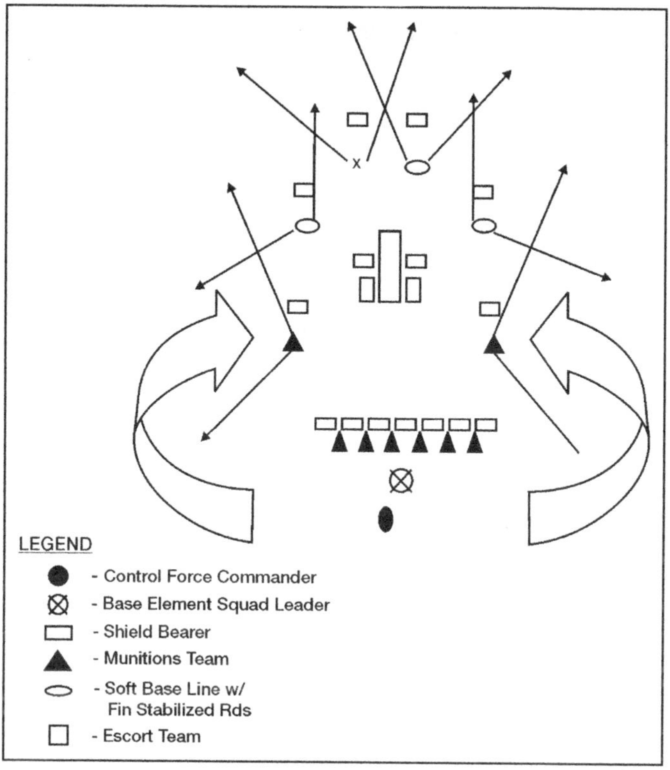

Figure V-21. Recovery of Downed Personnel or Stranded Vehicle (Big Johnson)

NOTE: In the event the crowd attacks personnel or a vehicle, the control force commander can call for a "Big Johnson." In a Big Johnson, the soft baseline deploys as usual except now the shield and shotgunners on the flank of the hard baseline also deploy creating a 190-degree perimeter around the personnel or vehicle. This maneuver must be executed quickly and with extreme violence of action because the crowd already has its hands on individuals in the force. The escort team deploys with the Big Johnson in order to escort or carry injured individuals. The shields without gunners provide protection for the escort team.

Chapter VI

LESSONS LEARNED

1. Background

Although lessons learned from employing NLW are still being studied and impacts analyzed, it is beneficial to look at the preliminary lessons gained from the NLW experience. These lessons are far from conclusive, because U.S. forces have yet to see the first large-scale employment of NLW. However, U.S. forces can gain tremendous insight from the experience of others and use what has been learned to guide future NLW employment.

2. Lessons Learned

a. Force Continuum. Using nonlethal options needs to be based on the principle of a graduated and proportional response, using the decisive force necessary to respond to the threat.

b. Attempts to legislate and regulate every aspect of NLW employment can prove disadvantageous to the on-scene commander. During the evacuation of U.N. forces in Somalia (OPERATION UNITED SHIELD, 1995), the ROE made an arbitrary and unnecessary distinction between situations justifying deadly force and those that justified only lesser means. Crowd control items were restricted to situations requiring deadly force. Many soldiers, airmen, and Marines asked, "why resort to nonlethal means at all if I have to wait until deadly force is authorized before I can shoot them with a beanbag?" Additionally, the use of sticky foam was limited to deadly force situations because of the perception that adversaries might suffocate from its application. This stipulation effectively eliminated sticky foam as a nonlethal option to the forces in Somalia.

c. With ROE that take into account the force continuum and encourage graduated response to nonlethal threats, the commander can employ NLW to enhance his flexibility, deterrence ability, and reaction to uncertain situations. To accomplish the mission with minimum loss of life and property, commanders need to have ROE that do not restrict them from properly employing NLW.

Lessons Learned from OPERATION JOINT GUARDIAN 1999 (Battalion Landing Team (BLT) 3/8 26 Marine Expeditionary Unit (MEU) Peace Operations in Kosovo:

ROE significantly impaired the employment of NLW. For example, OC was not authorized because it was considered a chemical agent. Nonlethal direct fire munitions were not authorized during the initial occupation of the province. This significantly hampered the individual Marine's ability to neutralize a threat where deadly force would not have been justified but action was required. Overly restrictive ROE thus negated the flexibility of response that NLW are supposed to provide. The intent of deploying NLW is to provide the operator the capability and flexibility to neutralize a threat without having to apply deadly force. Thus, the ROE for NLW should ideally be less restrictive, not the same or more restrictive, than those for lethal force.

It is often enticing for authors of ROE involving the use of NLW to base those ROE on the principle of minimum force necessary. This may be appropriate for base and station or homeland security ROE; however, for expeditionary operations the decision to mandate minimum force deserves critical consideration. Commanders in military operations have traditionally gained advantage over their enemy not only by the actual combat power that they bring to the fight, but also from that enemy's perceptions and beliefs about the intensity and resolve with which that combat power will be applied. When ROE are written stating that the confrontations will be addressed using the minimum force necessary, this advantage is lost. ROE developed for expeditionary nonlethal operations should be based on the concept of decisive and compelling force, rather than the minimum force. This will maintain an adversary's hesitancy to harass and endanger U.S. forces even after that adversary has tested or probed the response of those U.S. forces. Further, this will allow synergy to be gained through the complimentary employment of lethal and NLW by providing more response options to the small unit leader.

3. Employment

a. Through the use of modeling, simulation, and field exercises, new concepts for employing NLW are continually being developed. The force should train with NLW to the same degree as for lethal weapons. In this way, units become confident when using nonlethal options in chaotic and uncertain situations. The commander should push the decision to employ NLW to the lowest possible tactical level. Commanders who intend to use NLW in achieving military objectives must recognize the fluidity of the situations in which they are likely to be employed. The decisionmaking cycle is short, stressful, and the level of violence can move in both directions along the force continuum. In quick succession, a small unit leader might employ nonlethal munitions, decide to employ deadly force to an immediate specific threat that arises, then return to using nonlethal means in response to unarmed hostile elements.

b. A mixture of nonlethal "tools" has been determined to be advantageous in an operational scenario. As the situation develops, commanders can apply different nonlethal responses with respect to range/distance capability and severity of effect. For example, to disperse a crowd, area fire weapons would precede use of close-in point target nonlethal munitions.

c. An important point to remember is that NLW are just one element of a graduated response available to the force commander. Other factors make up a large contribution to the success of U.S. forces, including barrier plans and PSYOP. An example of this operation would be using barriers such as roadblocks or concertina wire, warning commands to inform a mob of potential responses based on their actions, and only then taking actions through escalated employment of NLW. These continued responses would involve smoke, RCA, and blunt munitions. A specific verbal warning should precede every escalation of nonlethal means.

d. Whenever NLW are employed, a separate lethal capability must always be in place. Troops must always be prepared to protect themselves as required by lethal means, because it is impossible to accurately predict the response of an individual or mob to NLW employment.

> **Lessons Learned from OPERATION ESSENTIAL HARVEST 2001 (U.S. Embassy Reinforcement operations conducted at the U.S. Embassy Skopje Macedonia by 5th Platoon, 2nd Fleet Antiterrorism Security Team [FAST] USMC):**
>
> The anticipated threats were violent demonstrations directed against U.S. facilities in the AO. In this case, USMC forces tasked to reinforce the embassy were equipped with a compliment of NLW. ROE took into account the capabilities and limitations of the NLW. The ROE and the NLW systems on the ground were coordinated and thus compatible. In addition, it was found that integrated employment of different nonlethal munitions greatly enhanced that force's ability to suppress the threats encountered (large rioting mobs). Random security patrols reinforced with military working dogs proved very effective. Reinforcement with multiple M203s provided USMC units with both a lethal and nonlethal capability in the same weapons system. The M203 was particularly useful as it gave the individual Marine the means to quickly transition between a lethal (5.56-mm) and nonlethal (40-mm nonlethal) capability. A basic understanding of the capabilities and limitations of nonlethal munitions enabled a relatively small force (40-50 Marines) to counter a sizable threat many times its size (500-1000 rioters). Although NLW were not used, the capability and preparation of the units provided the adequate deterrence that prevented the situation form escalating.

4. Publicity and Media

a. The commander can use the media to his advantage, as they will inform the local population of the military's capabilities and responses available. During OPERATION UNITED SHIELD in Somalia, the local media was informed of the Marines' nonlethal options and "new technologies" in order to discourage hostile actions. The media was encouraged to report this to the population. Intelligence had indicated the Somalis were planning to mix civilians with armed clansmen, creating hostile mobs. In fact, feuding warlords had previously successfully demonstrated this strategy on several occasions. Lack of deadly mob incidents implies Somalis were not anxious to test this new technology. Evidently, publicity had directly contributed to mission accomplishment.

b. The public dissemination of the presence of NLW capabilities must strike a balance between information sufficient to discourage hostile actions and information that enables the local population to learn too much of the capabilities of NLW. Too much information allows adversaries to develop countermeasures or avoid the NLW effects. Commanders who employ NLW systems must involve the PAO from the beginning of planning. The PAO must be familiar with the objectives of the mission, commander's intent, ROE, NLW and lethal weapons employed, and be prepared to advise the commander on the best methods of dealing with outside media. The commander should fully integrate the PA plan with PSYOP objectives and the concept of operations. Additionally, he must work the media to ensure the "sensationalism" of such a new technology does not create unrealistic expectations where he can be unfairly judged or constrained.

5. Capabilities and Impact on Mission Accomplishment

a. The commander should understand the unique capabilities of NLW and use them to accomplish the mission. He should use NLW in conjunction with the planned employment of the force and within established ROE to minimize loss of life and property while accomplishing the mission. The force requires early hands-on training to recognize the capabilities and limitations of NLW. Appendix D has sample training plans developed to meet these objectives.

b. The force commander must fully integrate planning for using nonlethal capabilities as quickly as possible in the staff planning cycle. In order to have a full range of alternatives for use against unarmed but hostile elements, the force needs to deliver nonlethal systems, complete necessary ROE and legal review, and make sufficient quantities of each NLW readily accessible to the force. The choice of nonlethal capabilities should be based on several factors and the choice can be made after addressing these questions:

(1) First, are the NLW systems available and can they be delivered on time?

(2) Second, are they compatible with, and do they complement existing weapons and training?

(3) Third, are there clear and unambiguous guidelines and ROE for employment?

(4) Finally, can the NLW be effectively employed to save lives and contribute to mission accomplishment?

c. Leaders need to anticipate local civilians developing effective countermeasures to their NLW and tactics. Therefore, prior to beginning their mission, leaders must determine and rehearse actions to take when nonlethal munitions are effectively countered.

d. Using nonlethal and lethal force are not mutually exclusive. In many instances, the commander can gain synergistic effects by using a combination of both. The following are some examples of this:

(1) Utilizing NLW to separate agitators from noncombatants with simultaneous employment of lethal force against hostiles or combatants where deadly force is justified.

(2) Utilizing NLW when making entry into structures or rooms to gain time in order to discern temper and intent of occupants prior to applying deadly force.

6. Knowing Your Adversary

a. Although commanders desire thorough intelligence of any potential adversary, such a goal may be difficult to reach in a NLW employment environment. During OPERATION UNITED SHIELD in Somalia and OPERATION UPHOLD DEMOCRACY (1995) in Haiti, a study of the local population's culture, history, and customs enabled commanders to anticipate and control many situations before they got out of hand. On the other hand, during OPERATION UPHOLD DEMOCRACY, poor intelligence and a lack of understanding between the two cultures led to the Haitians and Cubans being lodged together in Guantanamo Bay. This situation led to riots between factions who had a deep dislike for each other. By understanding a particular crowd's motivation, friendly forces may be better able to take necessary steps to prevent escalation resulting in the use of deadly force. Early training in crowd dynamics and riot control will enhance the effective use of NLW.

Lessons Learned from U.S. Peace Operations, 739th MP Battalion (Bn), January 2001

An Army MP platoon received a mission to clear a rioting crowd of 300 civilian locals from a main supply route. The platoon placed vehicle-mounted machine guns in overwatch positions to allow quick transition from NLW to conventional rounds. 12-gauge shotgun teams armed with fin stabilized rounds carried M9 pistols for lethal backup. Two volleys of shotgun and M203 nonlethal rounds successfully dispersed the crowd. The platoon had utilized INIWIC graduates to conduct preparatory training on using NLW. The NLW were extremely effective in pushing back the crowd and causing it to disperse. The 40-mm rubber ball round was moderately effective at close range, but very inaccurate at longer ranges. Shotguns were more effective than the M203 due to their high rate of fire and quick-loading capabilities.

b. Specific lessons learned during this operation are as follows:

(1) Placing interpreters is critical. The solution in this incident was to place interpreters approximately 25 meters behind formation so they could deploy forward as needed. Interpreters were often needed to communicate with crowds and with adjacent forces and coalition units that were organic to Kosovo forces (KFOR) that also responded to the incident.

(2) Leaders as well as soldiers should prepare to speak to the media. The command should have a PAO representative on hand to deal with the media.

(3) The commander should thoroughly brief the transition to lethal force from NLW.

(4) U.S. commanders in Kosovo engaged in the decisionmaking process had disparate levels of knowledge of nonlethal capabilities, constraints, and operational considerations. Therefore, leader training and orientation requires more emphasis.

(5) ROE were vague regarding the use of NLW. In many operations, NLW use is delegated to the on-site commander. Delegating to the lowest possible level is the preferred method.

(6) A unit needs a mixture of nonlethal rounds based on the size of crowd, composition of the crowd, available weapons systems, and synergistic effect of selected nonlethal combinations.

(7) When a RRF/QRF arrives on station with NLW to support a force equipped only with lethal means, the commander should address the following command and control concerns:

(a) Who makes the decision to employ NLW?

(b) Is the reinforced commander familiar with NLW?

(c) Who controls the reinforced unit?

(d) How does this relationship change as the situation develops?

7. Summary

a. The wide range of military operations in which U.S. forces are involved has demanded the introduction of NLW into the arsenal of weaponry available to commanders. NLW provide a graduated and proportional response to threats posed by a variety of adversaries. The ROE must be consistent with the rapidly changing situation of today's chaotic and uncertain military operations. This reality demands a high level of training in properly employing NLW.

b. Effective use of NLW involves a synergy between PSYOP activities, intelligence gathering, and professional media/PA plan. NLW effectiveness is enhanced through integration of these activities to reduce the chance adversaries will be able to develop effective countermeasures. NLW, when combined with lethal weapons, provide the commander with a force response across a wide range of military operations.

Appendix A

DOD Nonlethal Weapons and Equipment Item Description

The following is a cumulative list of NLW and equipment currently fielded within the DOD. Each service has various components of the following list in its inventory. A list of the service's items is located in Appendix E.

Figure A-1. Riot Face Shield

1. Riot Face Shield: The riot face shield is a non-ballistic face shield designed to protect the service member's face from objects thrown or attacks from non-ballistic weapons. (Figure A-1)

No picture available.

2. Ballistic Face Shield: The ballistic face shield provides the service member with face protection from a wide range of threats (i.e. debris, liquids, handthrown objects). The shield is manufactured from acrylic and bullet resistant materials to provide Level IIIA protection (9-mm and .44 magnum pistol rounds), weighs 3.4 pounds and is compatible with the PASGT helmet and M-17/M-40 protective masks.

Figure A-2. Riot Body Shield

3. Riot Body Shield: The riot body shield is a non-ballistic body shield designed to protect the service member's body from objects thrown or attacks from non-ballistic weapons. (Figure A-2)

No picture available.

4. Ballistic Body Shield with Light Kit: The ballistic shield provides the service member with torso protection from a wide range of threats (i.e. debris, liquids, hand thrown objects). The shield is manufactured from Spectra Shield resistant material painted anti-reflective flat black, and is capable of providing Level IIIA protection (9-mm and .44 magnum pistol rounds). The shield weighs 18 pounds and is 20 x 36 inches in size with a view port of 4 x 16 inches.

Figure A-3. Riot Shin Guard

5. Riot Shin Guard: The riot shin guard is a non-ballistic shin guard designed to protect the service member's shins from objects thrown or attacks from non-ballistic weapons. (Figure A-3)

No picture available.

6. Ballistic Shin Guards: Ballistic shin guards provide the service member with leg protection from a wide range of threats (i.e. debris, liquids, hand thrown objects). The guards are manufactured from Kevlar KM2 and provide Level IIIA protection (9-mm and .44 magnum pistol rounds). They are available in three sizes and weigh between 7 to 10 pounds.

Figure A-4. Expandable Riot Baton (with holster)

7. Expandable Riot Baton (with holster): The expandable riot baton provides the service member a nonlethal means of crowd control and self-defense. The baton comes with a mounting device that attaches the baton to a belt. The baton is 24 to 36 inches in length. (Figure A-4)

No picture available.

8. Wooden Baton: The wooden baton provides self-defense for the service member and the option between physical force and lethal force. The baton comes with a mounting device that attaches the baton to a belt. The baton is 36 inches in length.

No picture available.

9. Practice Riot Batons: The practice riot batons provide the service member with a safe dynamic training environment for properly using a riot baton. The batons are available in a variety of lengths.

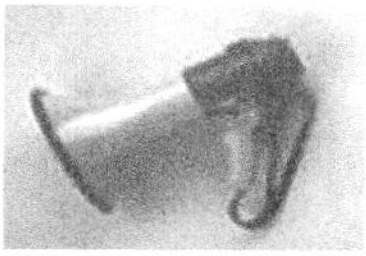

Figure A-5. Portable Bullhorn

10. Portable Bullhorn: The portable bullhorn provides the service member the capability to control his forces by projecting his voice above the noise and commotion created by crowds and mobs. (Figure A-5)

No picture available

11. Ground-Mounted Public Address System: The ground-mounted public address system is a communication enhancement device for conducting crowd control tactics. The system facilitates communication with the crowd in the absence of PSYOP support and assists with communication of commands to troops engaged in crowd-over-crowd noise.

No picture available.

12. M7 Individual Voice Amplification System: The M7 individual voice amplification system is a communication enhancement device for conducting crowd control tactics while wearing the M40 protective mask and using riot control agents. The M7 is fitted to the M40 protective mask to facilitate oral communication and increase the user's ability to communicate using radios and other devices.

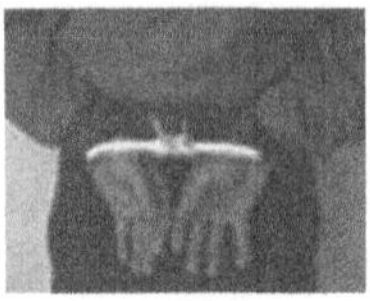 Figure A-6. Disposable Handcuff	13. Disposable Handcuff: The disposable handcuff provides service members with a means of restraining and controlling suspects, prisoners, and other individuals encountered during crowd control. The disposable handcuff is designed for one-time use. (Figure A-6)
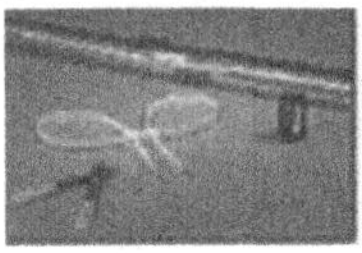 Figure A-7. Training Disposable Handcuff	14. Training Disposable Handcuff: The training disposable handcuff provides service members with a means for training in handcuffing procedures prior to actual use. The training disposable handcuff is red and reusable. (Figure A-7)
 Figure A-8. MK-4 Pepper Spray Pouch	15. MK-4 Pepper Spray Pouch: The MK-4 pepper spray pouch is designed to carry the MK-4 pepper spray canister. (Figure A-8)
 Figure A-9. MK-4 Individual Live OC Pepper Spray	16. MK-4 Individual Live OC Pepper Spray: The MK-4 individual pepper spray has a range of up to 12 feet in a target-specific stream. It contains 30 to 35 half-second bursts. (Figure A-9)
No picture available.	17. M-36 Individual Riot Control Disperser: M-36 Individual Riot Control Disperser has a range of up to 12 feet in a target-specific stream. It contains 30 to 35 half-second bursts. It is loaded with either CN or CS agent.
 Figure A-10. MK-9 Pepper Spray Pouch	18. MK-9 Pepper Spray Pouch: The MK-9 pepper spray pouch is designed to carry the MK-9 pepper spray canister. (Figure A-10)

Figure A-11. MK-9 Squad Live OC Pepper Spray

19. MK-9 Squad Live OC Pepper Spray: The MK-9 squad pepper spray has a range of up to 15 feet in a target-specific fog. It contains 8 to10 half-second high volume bursts. (Figure A-11)

Figure A-12. MK-46 Platoon Live (Horizontal) OC Pepper Spray

20. MK-46 Platoon Live (Horizontal) OC Pepper Spray: The MK-46 platoon pepper spray has a range of up to 25 to 30 feet in a target-specific dispersed spray pattern. It contains up to 12 one-second high volume bursts. (Figure A-12)

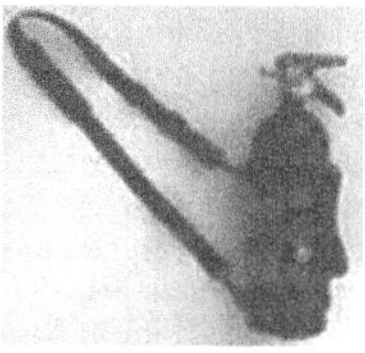

Figure A-13. MK-46 Sling

21. MK-46 Sling: The MK-46 sling gives a service member the ability to sling the MK-46 canister on the shoulder. (Figure A-13)

Figure A-14. MK-46 Hardware Kit

22. MK-46 Hardware Kit: The MK-46 hardware kit consists of a transfer tube, funnel, and transfer bottle. A compressed nitrogen source is also required to utilize the kit. (Figure A-14)

Figure A-15. MK-46 Refill Live OC Pepper Spray

23. MK-46 Refill Live OC Pepper Spray: The MK-46 refill solution comes in a one-gallon container. (Figure A-15)

Figure A-16. MK-4 Individual Inert OC Pepper Spray

24. MK-4 Individual Inert OC Pepper Spray: The MK-4 inert individual pepper spray has a range of up to 12 feet in a target-specific stream. It contains 30 to 35 half-second bursts. It provides the service member with an inert canister for familiarization and training purposes. The inert device is loaded with a non-irritant, non-toxic formulation payload. The device may also contain a non-toxic, water-soluble marker to allow immediate feedback on content delivery to target. The MK-4 device is marked "Inert." (Figure A-16)

Figure A-17. MK-9 Squad Inert OC Pepper Spray

25. MK-9 Squad Inert OC Pepper Spray: The MK-9 squad inert pepper spray has a range of up to 15 feet in a target-specific fog. It contains 8 to 10 half-second high volume bursts. It provides the service member with an inert canister for familiarization and training purposes. The inert device is loaded with a non-irritant, non-toxic formulation payload. The device may also contain a non-toxic, water-soluble marker to allow immediate feedback on content delivery to target. The inert MK-9 device is prominently marked "Inert" with a yellow sticker attached. (Figure A-17)

Figure A-18. MK-46 Platoon Inert OC Pepper Spray

26. MK-46 Platoon Inert OC Pepper Spray: The MK-46 platoon inert pepper spray has a range of up to 25 to 30 feet in a target-specific dispersed spray pattern. It contains up to 12 one-second high volume bursts. It provides the service member with an inert canister for familiarization and training purposes. The inert device is loaded win a non-irritant, non-toxic formulation payload. The device may also contain a non-toxic, water-soluble marker to allow immediate feedback on content delivery to target. The inert MK-46 device is prominently marked "Inert" with a yellow sticker attached. (Figure A-18)

Figure A-19. MK-46 Refill Inert OC Pepper Spray

27. MK-46 Refill Inert OC Pepper Spray: The MK-46 refill solution comes in a one-gallon container. (Figure A-19)

Figure A-20. FIST Training Suit

28. FIST Training Suit: The FIST training suit is made of closed-cell shock-absorbent foam that allows service members to hone their riot baton skills through realistic hand-to-hand engagement training. The FIST suit can absorb blows inflicted by the current expandable and wood riot baton. It provides protection for the head, face, hands, and legs, without significant degradation of the wearer's mobility. The FIST suit consists of a helmet, chest and back protector, bicep and forearm protectors, groin and buttock protectors, thigh and shin protectors, gloves, and carry bag. (Figure A-20)

NOTE: FIST is the current manufacturer of this training suit).

Figure A-21. Riot Training Strike Bag

29. Riot Training Strike Bag: The riot training strike bag gives the service member a strike bag for training and proficiency in open-hand control techniques. (Figure A-21)

Figure A-22. Caltrop

30. Caltrop: The caltrop is a field expedient area denial system. Securing approximately 50 caltrops to 550 military cord best employs it. This allows a service member to quickly toss/recover the system, using the caltrops to deny vehicle/pedestrian movement in a designated area. The four prongc are approximately two inches tall and will puncture tires or boots. (Figure A-22)

Figure A-23. Road Side Spike Strip

31. Road Side Spike Strip: The spike strip is designed to puncture a pneumatic tire as the vehicle makes contact with the spike strip. The spikes pivot into the tire and the tip-guards are pushed down. The spikes are imbedded deeply into the tire. The spikes are pulled out of the unit and remain in the tire. On the second revolution they will be pushed directly into the tire, allowing air to escape slowly without causing a blowout or sudden loss of control. The average tire will pick up 4 to 6 spikes with the deflation time of about 12 to 20 seconds. It allows a service member to quickly toss/recover the system, impeding vehicle movement in a designated area. It allows a force to provide additional physical security in areas where the caltrop cannot be tactically employed. (Figure A-23)

Figure A-24. High-Intensity Search Light

32. The high-intensity searchlight is a rugged, lightweight, hand-held, high-intensity spotlight for use during low-light conditions. Two different high-intensity lights were fielded. The Maxabeam light (shown on top) was fielded in 1997. It was replaced by the Nighthunter light (shown underneath) in 2000. (Figure A-24)

Figure A-25. Individual Light Holster (6P)

33. Individual Light Holster (6P): The 6P individual light holster is designed to hold the 6P individual light. (Figure A-25)

Figure A-26. 6P Individual Light

34. 6P Individual Light: The 6P individual light is issued to service members within the platoon. (Figure A-26)

Figure A-27. 6P Individual Light Accessory Kit

35. 6P Individual Light Accessory Kit: The 6P individual light accessory kit includes a spare light bulb installation kit. (Figure A-27)

Figure A-28. 3V Lithium Batteries

36. 3V Lithium Batteries: Two 3V Lithium Batteries are required to operate the 6P individual light. (Figure A-28)

Figure A-29. 12-Gauge Buttstock Cuff

37. 12-Gauge Buttstock Cuff: The buttstock cuff is designed to fasten to the buttstock of most shotguns to hold seven 12-gauge cartridges for quick loading of the shotgun. (Figure A-29)

Figure A-30. 12-Gauge Utility Pouch

38. 12-Gauge Utility Pouch: The 12-gauge utility pouch is designed to hold twenty-five 12-gauge munitions and attaches to the service member's combat gear. (Figure A-30)

Figure A-31. 40-mm Carrying Pouch

39. 40-mm Carrying Pouch: The 40-mm carrying pouch is partitioned in a manner to provide the service member with a means of carrying not less than six 40-mm nonlethal rounds in separate compartments. It attaches to the service member's combat gear. (Figure A-31)

Figure A-32. Diversionary/Rubber Ball Grenade Pouch

40. Diversionary/Rubber Ball Grenade Pouch: The diversionary/rubber ball grenade pouch is an ammunition carrying pouch partitioned in a manner to provide the service member with a means of holding up to six rubber ball grenades or six flash bang grenades in separate compartments. (Figure A-32)

Figure A-33. 12-Gauge Mossberg 500A2/590 Shotgun

41. 12-Gauge Mossberg 500A2/590 Shotgun: The 12-gauge Mossberg 500A2 or 590 shotgun are the only 12-gauge shotguns currently authorized for firing 12-gauge nonlethal projectile munitions. The launching cup for the rubber ball grenade will attach to the Mossberg 500A2 Shotgun. The launch cup will not attach to the Mossberg 590 shotgun. Until the Joint Service Combat shotgun can effectively fire nonlethal 12-gauge rounds, the 12-gauge Mossberg 500A2 or 590 shotgun configuration meets the requirement of the NLW force. (Figure A-33)

Figure A-34. 12-Gauge M870 (modified) Version 2 (USCG)

42. 12-Gauge M870 (modified) Version 2 (USCG): The modified Remington 870 shotgun has a fixed synthetic stock and pistol grip. The pistol grip gives the shooter more control of the weapon and helps dissipate recoil from the shoulder. The synthetic materials used are more durable and lighter weight than the wood stocks used on the Standard M870. The barrel is fitted with a rifled choke tube for firing slugs (solid projectiles) to increase accuracy. It is also equipped with an Advanced Combat Optical Gunsight that was developed for use in "both-eyes-open" close quarters battle. The advantage is rapid target acquisition without the need to align front and rear sights. Batteries do not power the illumination sight; it contains tritium gas.

In the Coast Guard, a yellow "stock sock" will be used whenever the shotgun is loaded with nonlethal munitions. This removable nylon sleeve is used to identify the weapon as a nonlethal delivery system. (Figure A-34)

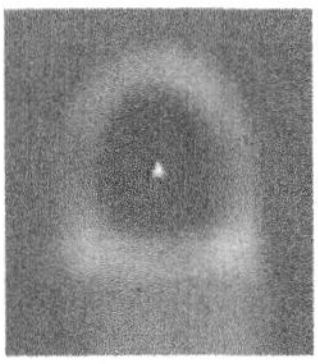

Figure A-35. Advanced Combat Optical Gunsight (ACOG) Reflex Sight (USCG)

43. Advanced Combat Optical Gunsight (ACOG) Reflex Sight (USCG): Modified shotguns are equipped with this site. (Figure A-34)

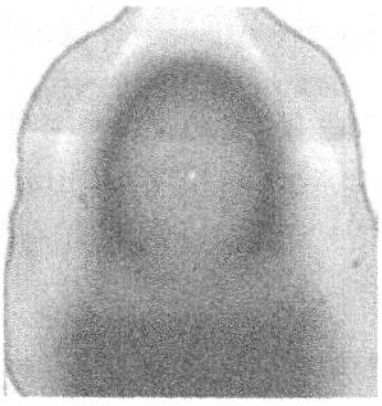

Figure A-36. ACOG Reflex Sight (USCG)

44. ACOG Reflex Sight (USCG): The dual-illuminated, Trijicon® Reflex sight was developed for the military for use in both-eyes-open close quarters battle (CQB). The advantage is quick target acquisition thus eliminating the need to align front and rear sights. Batteries do not power the illumination sight; it contains tritium gas. (Figure A-35)

Equipment Currently Being Fielded

1. 40-mm Crowd Dispersal Cartridge (CDC):

This direct fire, low-hazard, non-shrapnel-producing, multiple ball projectile provides friendly forces the capability to conduct crowd control. The CDC uses nonlethal, blunt trauma to stop, confuse, disorient, and/or temporarily incapacitate personnel and increase standoff distance between friendly forces and the hostile crowd. Primarily designed as an area effect munition, the CDC can be used on individuals. The M203 40-mm grenade launcher fires the CDC and contains 48, .48 caliber rubber balls. Effective range is 10 to 30 meters. (Figures A-37 and A-38)

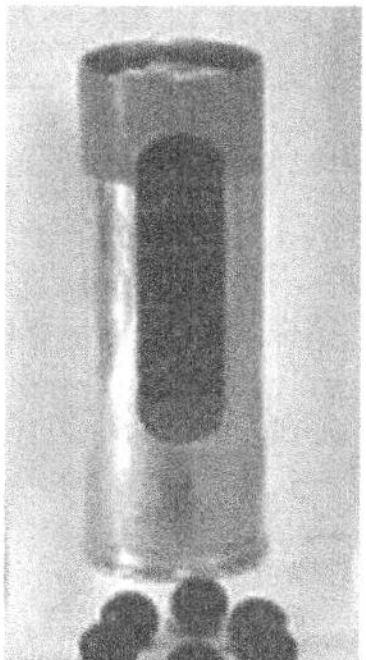

Figure A-37. Crowd Dispersal Cartridge

Figure A-38. Soldier Employing CDC with M203

2. Portable Vehicle Arresting Barrier (PVAB):

PVAB is a lightweight, portable, manually emplaced and recoverable barrier entrapment device to create, on command, a barrier that stops vehicles driven by known or suspected hostile forces with minimal damage to the vehicle or its occupants. It is used at dismount points, checkpoints, and roadblocks to prevent unauthorized access by wheeled vehicles into or out of areas (e.g. airfields, ports, command posts [CP]). Two individuals can easily emplace it, in approximately four hours. In stand-by mode, PVAB folds down into something similar to a speed bump (vehicles pass freely). In capture mode, the net springs up on two hydraulic pistons, to maximum height in less than a second. PVAB can be set up to span 12 to 24 feet and can stop a light truck (up to 7,500 pounds) at speeds up to 45 miles per hour, with a stopping distance of less than 200 feet PVAB can be containerized: 53.75 cubic feet (7 boxes, 4 large-3 small)—1,044 pounds on approximately one pallet. (Figures A-39 and A-40)

Figure A-39. PVAB in Action (Far View)

Figure A-40. PVAB in Action Near View

3. Modified Crowd Control Munition—Ground Emplaced (MCCM-GE):

The MCCM-GE (Figure A-41) is a nonlethal munition, housed in a casing similar to the current claymore mine, which will stop, confuse, disorient and/or temporarily incapacitate area targets/personnel at close range. The force can use it to provide crowd control by dispersing hostile crowds and force protection by integrating it into a layered perimeter defense. It delivers 600, .32 caliber rubber balls as well as a flash bang effect. The MCCM-GE gives the commander or small unit leader greater flexibility to apply nonlethal force as a first line of defense against hostile noncombatants. Effective range is 5 to 15 meters. (Figure A-42)

Figure A-41. Modified Crowd Control Munition-Ground Emplaced (MCCM-GE)

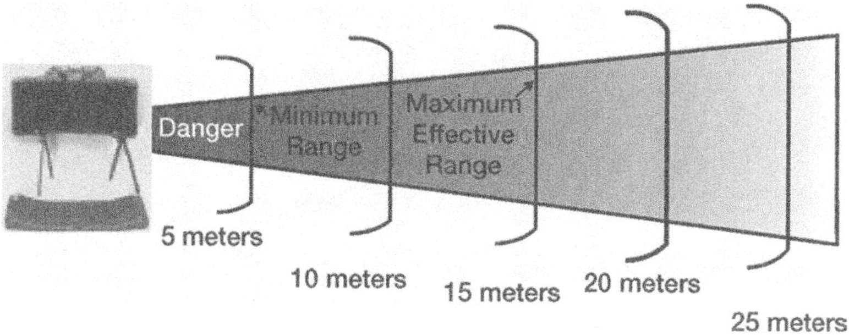

Figure A-42. MCCM-GE Range Chart

4. 66-mm Vehicle Mounted Nonlethal Grenade Launcher (VMNLGL):

The VMNLGL is a nonlethal grenade-launching device mounted on high mobility multipurpose wheeled vehicle (HMMWV). It is an indirect fire-support system that can deliver a 66-mm distraction device (flash-bang), CS, or blunt trauma (sting ball) payload. The distraction grenade uses a pyrotechnic charge for crowd control through audio and visual stimuli. The blunt trauma grenade uses a pyrotechnic charge to discharge 140, .32 caliber (cal.), rubber balls to achieve crowd control through audio, visual, and physical stimuli. The force uses VMNLGL to enhance the ability of friendly forces to conduct force protection without direct contact with hostile or threatening crowds. It enforces standoff distances and deters potential threats. It can disperse or subdue crowds, keep personnel from surrounding or climbing on patrol vehicles, slow or stop threatening personnel or vehicles by confusing, disorienting, and/or temporarily incapacitating individuals for capture. The grenade is fired from the grenade launchers traditionally found on tracked vehicles. Vehicle-launched grenades have three submunitions designed to ground burst. Effective range is 50 to 100 meters. An adjustable bracket allows engagement at ranges of 50 meters, 75 meters, and 100 meters. (Figure A-44)

Figure A-43. 66-mm Vehicle Mounted Nonlethal Grenade Launcher (VMNLGL)

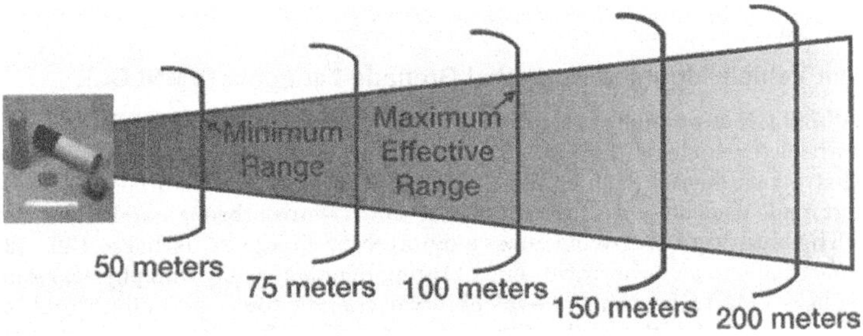

Figure A-44. VMNLGL Range Chart

5. Running Gear Entanglement System (RGES)

The Running Gear Entanglement System (RGES) is series of short, separate, looped polyethylene lines that are attached to a single, long polyethylene base line. This entanglement net fouls the propellers of planing hull small boats. The capability to manually deploy the net from a soft-sided bag has been fielded for USCG use in limited numbers. The USCG is in the process of incorporating three other variants of RGES into its inventory of nonlethal capabilities. These include aerial-delivered RGES, surface launched RGES to deploy from vessels, and static barrier RGES. The static barrier RGES has broad application into support homeland security, physical security and force protection missions. The surface launched RGES has direct application in the USCG's law enforcement mission wherever the requirement exists to stop a non-compliant vessel. Both of these systems are expected to be fielded in the fall/winter of CY02. The air launched version is currently still in the concept development phase and is not yet fully developed. (Figure A-45)

Figure A-45. Running Gear Entanglement System (RGES)

a. RGES Surface Delivery System: The USCG can use stern-mounted, manually deployed RGES from a cutter's small boat to stop slow-moving migrant-type craft. (Figure A-46)

(1) Uses COTS compressed air cylinder launcher.

(2) Removable, fix-mounted launcher on side of craft.

(3) Launch projects device ahead of target.

Figure A-46. RGES Surface Delivery System

b. RGES Barrier

(1) Provides static barrier around ship or other waterside asset.

(2) Provides a visible demarcation line, establishes zones, deters intrusions. (Figure A-47)

Figure A-47. RGES Barrier Employment

c. RGES Aerial Delivery System (ADS): RGES is aerial deployed from the ADS ahead of target aboard a H60 helicopter (Figure A-48). Device should float in the water ahead of target ensuring target runs over device. Device fouls propeller or rudder until target stops and awaits boarding.

(1) Uses COTS compressed air cylinder launcher

(2) Removable, fix-mounted launcher on helicopter

(3) Launch projects device ahead of target. (Figures A-49)

Figure A-48. RGES Aerial Delivery System (ADS)

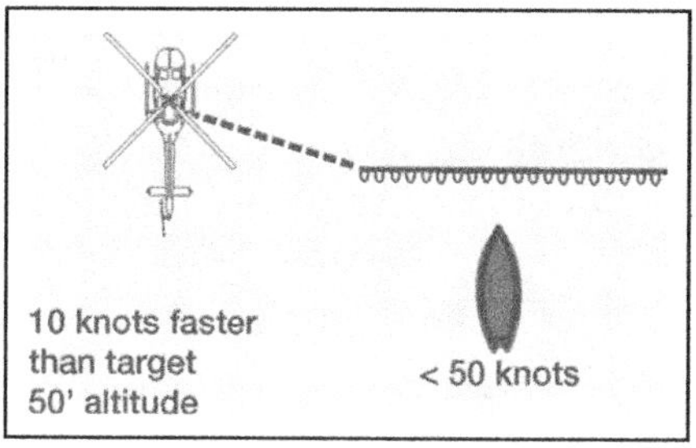

Figure A-49. RGES Aerial Delivery System (ADS)

6. Warp Cutter

a. Warp cutting is a nonlethal use of force, which has been safely and effectively employed by several countries to enforce fisheries laws in their respective exclusive economic zones (EEZ). The warp cutter is a device that closely resembles a grappling hook or small anchor. It has hardened steel blades attached to the crux of the grapnels. It is towed behind an enforcement vessel and across the warps of a non-compliant fishing vessel. As the cutting device engages the warps, it slices through them, separating the net from the fishing vessel.

b. The USCG uses the warp cutter to compel foreign vessels suspected of illegally fishing in the U.S. EEZ to comply with lawful orders to stop and be boarded. The USCG may cut the trawl warps of non-compliant vessels as a means of compelling them to stop for a boarding.

Figure A-50. Warp Cutter

Appendix B

Summary: Specialty Impact Munitions

12-Gauge Launch Cartridge

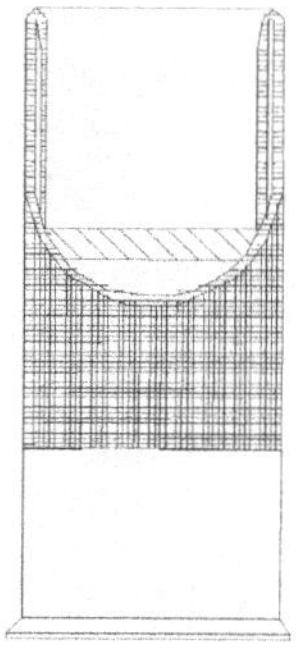

Using Services: USN, USMC
Status: Fielded 1998
DODIC: AA30
NSN: 1305-01-464-8389
Acquisition Activity: MARCORSYCOM—NSWC Crane, IN
Further Information:

Commercial: MARCORSYSCOM (703) 784-9395
NSWC Crane (812) 854-5801

Color .... Clear case
Diameter 12-gauge
Projectile .... N/A
Length .... 2.00 inches (51 mm)
Total Weight........................... 13.6 grams
Propellant Weight................. 46.3 grains (black powder)
Muzzle Velocity...................... 100 feet/second (30.5 m/sec)—rubber ball grenade
Acceptance Accuracy.......... 200 feet (61 m) + 25 feet (7.6 m) at 30-degree launch angle using rubber ball grenade and 12-gauge launch cup

Projectile Specification: N/A

- Use 12-gauge launch round in combination with rubber ball grenade and 12-gauge launch cup to hurl a grenade a distance of 200 feet (61 m) + 25 feet (7.6 m) when launch angle is approximately 30 degrees.
- Chamber launch cartridge only after installing grenade in launch cup and pulling safety pin from grenade.

Figure B-1. 12-Gauge Launch Cartridge

12-Gauge Bean Bag Cartridge

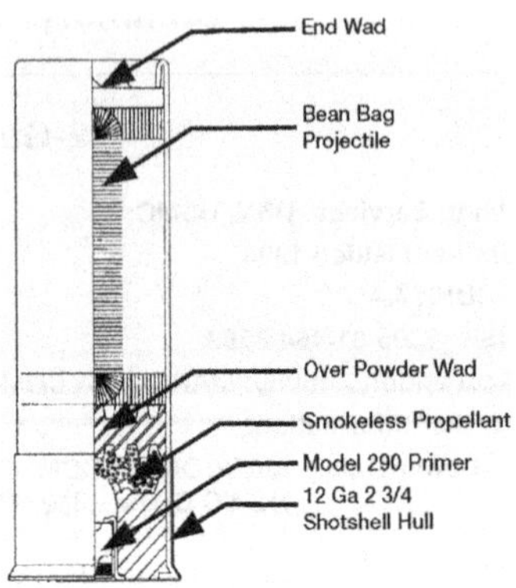

Using Services: USN, USMC

DODIC: AA29

NSN: 1305-01-454-0191

Status: Fielded 1998

Acquisition Activity: MARCORSYCOM—NSWC Crane, IN

Further Information:

Commercial: MARCORSYSCOM (703) 784-9395
NSWC Crane (812) 854-5801

Color .... Clear or red case
Diameter .... 12 gauge
Projectile .... 1 bean bag
Length .... 2.45 inches (62 mm)
Total Weight .... 49.9 grams
Propellant Weight .... 3.4 grains
Muzzle Velocity .... 300 + 25 feet/second (91.4 + 7.62 m/sec)
Acceptance Accuracy .... 2 feet (.6 m) X 6 feet (1.8 m) at 100 feet (30.3 m) with Mossberg 500A2

Shotgun Projectile Specification:

Shape .... Square bag or tube sock
Material .... Nylon or cotton-polyester
Filler .... #9 lead shot
Filler weight .... 44.4 grams (maximum)

- Fatal injuries are possible at employment distances of less than 20 feet (6 m)
- At distances of 20 feet (6 m) to 40 feet (12 m), target area should be center mass
- Training needs to reinforce that head shots are NOT ACCEPTABLE
- Do not skip fire this round! Round becomes unpredictable upon striking the ground
- Round is most effective against individually selected targets (point round)
- End wad has low light marker

Figure B-2. 12-Gauge Bean Bag Cartridge

12-Gauge Rubber Fin Stabilized Cartridge

Using Services: USN, USMC, USCG

Status: Fielded 1998

DODIC: AA31

NSN: 1305-01-454-0189

Acquisition Activity: MARCORSYCOM—NSWC Crane, IN

Further Information:

Commercial: MARCORSYSCOM (703) 784-9395
NSWC Crane (812) 854-5801

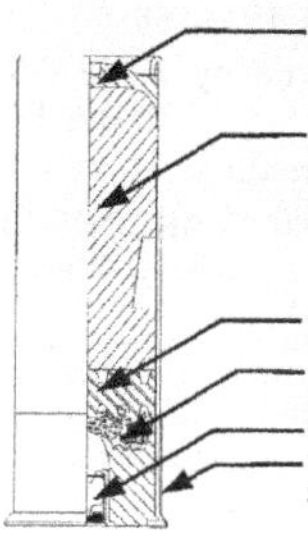

Color ....Clear or red case
Diameter ....12 gauge
Projectile ....1 rubber projectile
Length ....2.45 inches (62 mm)
Total Weigh ....9.1 grams
Propellant Weight.................4.8 grains
Muzzle Velocity.....................530 + 65 feet/second
(161.5 + 19.8 m/sec)
Acceptance Accuracy..........2 feet (.6 m) X 6 feet (1.8 m) at 100 feet (30.3 m) with Mossberg 500 A2 shotgun
Projectile Specification: ......Caliber 12 gauge
Material ....Rubber
Hardness ....Durometer 70-80 Shore "A" scale
Weight ....7.8 grams
Temperature Limits (demonstrated):
Upper limit...125° F
Lower limit...0° F

Shipping Data: Hazard classification/division and storage compatibility group...1.4S
DOT Class ....ORM-D
DOT Markings Cartridge......Small Arms
UNO serial number...............0012

- Fatal injuries are possible at employment distances of less than 20 feet (6 m); USCG: 15 yards
- At distances of 20 feet (6 m) to 40 feet (12 m), target area should be center mass; USCG: 15 to 30 yards
- Training needs to reinforce that head shots are NOT ACCEPTABLE
- Do not skip fire this round! Round becomes unpredictable upon striking the ground
- Round is most effective against individually selected targets (point round)
- End wad has low light marking

Figure B-3. 12-Gauge Rubber Fin Stabilized Cartridge

40-mm Foam Rubber Baton Cartridge

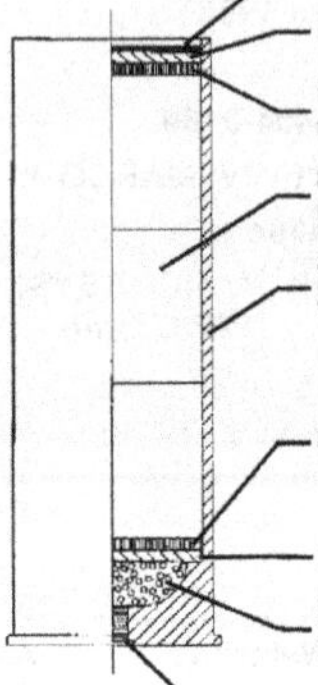

Using Services: USN, USMC

Status: Fielded 1998

DODIC: BA07

NSN: 1310-01-453-9168

Acquisition Activity: MARCORSYSCOM—NSWC Crane, IN

Further Information:

Commercial: MARCORSYSCOM (703) 784-9395
NSWC Crane (812) 854-5801

Color......................................Aluminum (case)
Diameter................................40 millimeter
Projectile ....3 foam batons
Length ....4.78 inches (121.4 mm)
Total Weight...........................201 grams
Propellant Weight.................13.0 grains
Muzzle Velocity......................325 + 35 feet/second (99 + 10.6 m/sec)
Acceptance Accuracy..........6 feet (1.8 m) X 3 feet (.9 m) at 50 feet (15.2 m) with M203

Projectile Specifications:

Number............................3
Material............................Foam rubber
Hardness ....Durometer 25 + 10 on Shore "A" scale
Weight..............................250 to 280 grains

- Fatal injuries are possible at employment distances of less than 20 feet (6 m)
- At distances of 20 feet (6 m) to 40 feet (12 m), target area should be center mass
- Training needs to reinforce that head shots are NOT ACCEPTABLE
- Round is most effective against point targets (individual personnel)
- End wad has low light marker

Figure B-4. 40-mm Foam Rubber Baton Cartridge

40-mm Rubber Ball Cartridge

Using Services: USN, USMC
Status: Fielded 1998
DODIC: BA08
NSN: 1310-01-453-9154
Acquisition Activity: MARCORSYSCOM—NSWC
Crane, IN
Further Information:
Commercial: MARCORSYSCOM (703) 784-9395

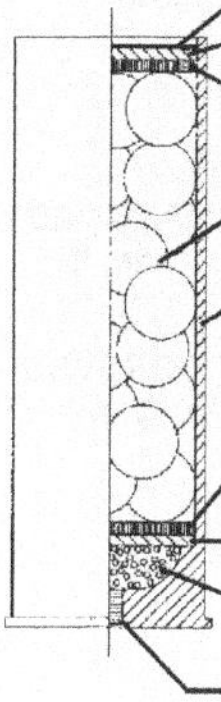

Color .... Aluminum (case)
Diameter .... 40 millimeter
Projectile .... 24 rubber balls
Length .... 4.78 inches (121.4 mm)
Total Weight .... 201 grams
Propellant Weight .... 7.3 grains
Muzzle Velocity .... 325 + 35 feet/second
(99 + 10.6 m/sec)
Acceptance Accuracy .... 6 feet (1.8 m) x 9 feet (2.7 m) at
50 feet (15.2 m) with
M203NSWC Crane
(812) 854-5801

Projectile Specifications:
Caliber 60
Number .... 22 (minimum)
Material .... Rubber
Hardness .... Durometer 60 + 10 on Shore "A" scale
Weight .... 2.2 grams each

- Fatal injuries are possible at employment distances of less than 20 feet (6 m)
- At distances of 20 feet (6 m) to 40 feet (12 m), target area should be center mass
- Training needs to reinforce that head shots are NOT ACCEPTABLE
- Round is most effective against area targets (multiple personnel)
- End wad has low light marker

Figure B-5. 40-mm Rubber Ball Cartridge

Rubber Ball Grenade

Using Services: USN, USMC

Status: Fielded 1998

DODIC: GG04

NSN: 1310-01-454-0132

Acquisition Activity: MARCORSYSCOM—NSWC Crane, IN

Further Information:

Commercial: MARCORSYSCOM (703) 784-9395
NSWC Crane (812) 854-5801

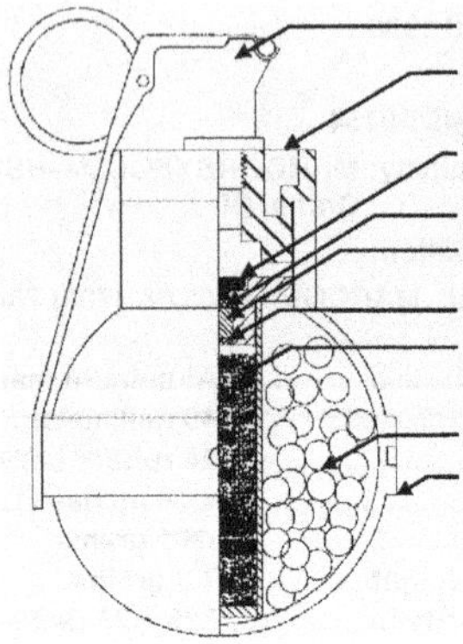

Color ....Black with white lettering
Diameter ....3.12 inches (79 mm)
Projectiles ....100 rubber balls (minimum)
Height ....5.25 inches (133.4 mm)
Total Weight..........................290 grams
Explosive Weight..................59 grams
Total Time Delay...................3.0 + 0.5 seconds

Employment Distance:

Hand Thrown ....50 feet (15.2 m)
Shotgun Launched ....200 feet (61 m)

Projectile Specifications:

Number................................100 (minimum)
Material................................Rubber
Hardness.............................Durometer 70 to 80 on Shore "A" scale
Diameter..............................0.25 inches (6.4 mm) (minimum)

- Fuze separates from the body 1.3 to 1.7 seconds after primer ignition.
- Second delay causes burster charge detonation 3.0 + 0.5 seconds average after primer ignition. Rubber balls are then dispersed in a 50-feet (15-m) circular pattern.
- 12-gauge shotgun with launch cup and launch cartridge can launch grenade 200 feet (61 m) + 25 feet (7.6 m)

Figure B-6. Rubber Ball Grenade

Practice Grenade Body

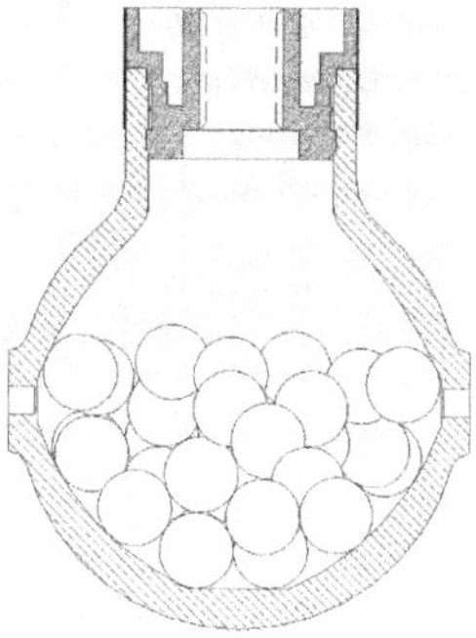

Using Services: USMC

Status: Fielded 2002

DODIC: GG05

NSN: Pending

Acquisition Activity: MARCORSYSCOM – NSWC Crane, IN

Further Information:

Commercial: MARCORSYSCOM (703) 784-9395
NSWC Crane (812) 854-5801

Diameter................................3.12 inches (79 mm)
Projectiles............................N/A
Height....................................4.2 inches (106 mm)
Total Weight..........................208 grams
Explosive Weight..................N/A (inert)
Total Time Delay..................3.0 + 0.5 seconds when used with practice grenade fuze

Employment Distance:

Hand Thrown..................50 feet (15.2 m)
Shotgun Launched.........200 feet (61 m) using Mossberg launch cup and 12-gauge launch cartridge (AA30).

- Practice grenade body is reusable up to 25 times and is a component of a practice grenade assembly, consisting of the body and a practice grenade fuze.
- Practice grenade assembly accurately replicates the GG04 rubber ball grenade and is used for training purposes.

Figure B-7. Practice Grenade Body

Practice Grenade Fuze

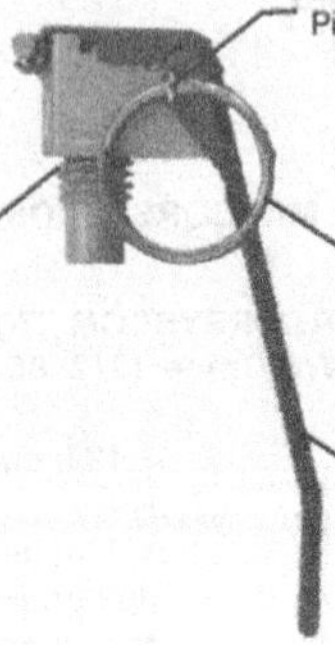

Using Services: USMC

Status: Fielded 2002

DODIC: Pending

NSN: Pending

Acquisition Activity: MARCORSYSCOM—NSWC Crane, IN

Further Information:

Commercial: MARCORSYSCOM (703) 784-9395
NSWC Crane (812) 854-5801

Color ....Blue safety lever with brown stripe
Projectiles ....N/A
Height ....4.25 inches (107 mm)
Total Weight...........................82 grams
Explosive Weight...................Not available
Total Time Delay...................3.0 + 0.5 seconds

Employment Distance when used with GG05 Practice Grenade Body:
Hand Thrown ....50 feet (15.2 m)
Shotgun Launched...........200 feet (61 m)—using Mossberg launch cup and 12-gauge launch cartridge (AA30).

- Employ practice grenade fuze with the practice grenade body to create a practice grenade assembly consisting of the body and a practice grenade fuze.
- Practice grenade assembly accurately replicates the GG04 rubber ball grenade and is used for training purposes.

Figure B-8. Practice Grenade Fuze

Nonlethal Grenade Launch Cup

Using Service: USMC

Status: Fielded 2002

DODIC: N/A

NSN: Pending

Acquisition Activity: MARCORSYSCOM—NSWC Crane, IN

Further Information: Commercial: MARCORSYSCOM
(703) 784-2006 Ext. 2733
NSWC Crane (812) 854-5801

NOTE: Cup body shown is representative of cup bodies used on all shotguns. Adapter shown is subject to change depending on the shotgun used and design improvements.

Color .... Black
Diameter (internal) .... 3.12 inches (79 mm)
Length (without adapter) ..... 4 inches (102 mm)
Weight .... 397 grams (maximum)

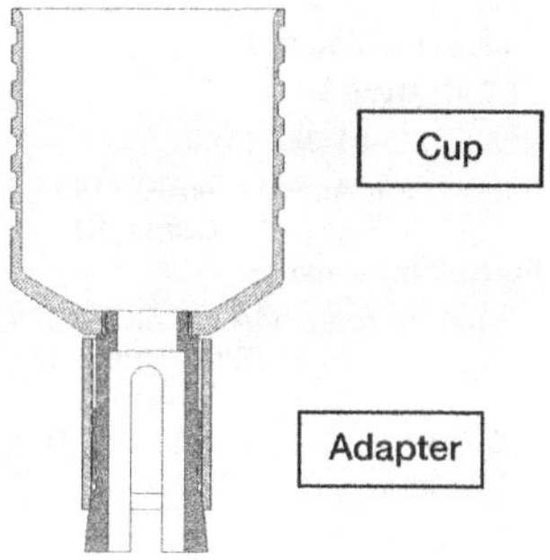

Cross Section View

- Use launch cup in concert with AA30 launch cartridge and either the GG04 rubber ball grenade or GG05 practice grenade assembly
- Attach launch cup and remove from the shotgun without using tools
- Launch cup can withstand accidental detonation of rubber ball grenade without fragmenting

Figure B-9. Nonlethal Grenade Launch Cup

Diversionary Charge, MK 141, MOD 0

Using Services: USN, USMC
Status: Fielded 1997
DODIC: DWBS
NSN: 1375-01-387-5748
Acquisition Activity: MARCORSYSCOM—NSWC Crane, IN
Further Information:
Commercial: MARCORSYSCOM (703) 784-9395
NSWC Crane (812) 854-3934

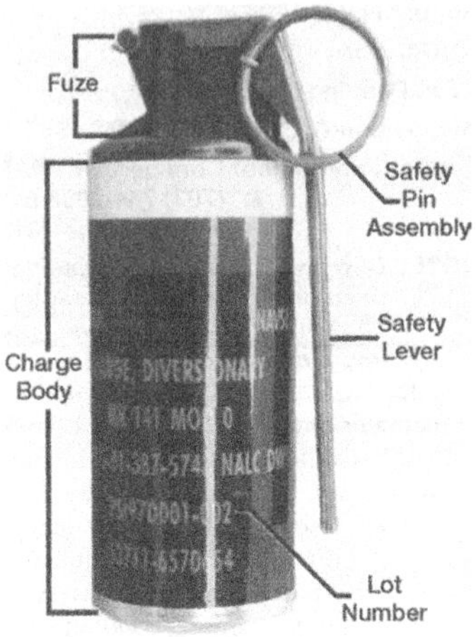

Color	Brown with white stripe
Diameter	1.77 inches (45 mm)
Projectile	N/A
Length	5.214 inches (132.4 mm)
Total Weight	99.8 grams
Net Explosive Weight	17.8 grains
Light Output	2 million candlepower for 4 milliseconds
Noise Output	176 to 185 decibels
Material	Molded and machined polyurethane
Time Delay:	
Initial	1.26 seconds (minimum)
Total	1.34 to 1.84 seconds

- Diversionary charge can cause serious injury if held at time of detonation
- Light and noise most effective in confined space to stun individuals
- Most often employed for hostage rescue

Figure B-10. Diversionary Charge, MK 141, MOD 0

Cartridge, 12-Gauge Rubber Sting Ball™

Using Services: USCG

Status: Fielded

DODIC: N/A Model No. 2552

NSN: Commercial Procurement

Acquisitions Activity: Commandant (G-OCU)

Further Information: Commandant (G-OCU) (202) 267-1522

Model Number	31-caliber sting-ball
Type	Multiple projectile rubber pellets
Range	10 yards (9 m)
Projectiles	18 (average)
Cartridge Material	Plastic
Projectile Weight	.35 grams each
Projectile Material	.31-caliber rubber pellets
Overall Height	2.75 inches (70 mm)
Overall Weight	18 grams
Compatibility	12-gauge shotguns
Velocity	400 feet/second

USCG

- Fatal injuries are possible at employment distances of less than 5 yards.
- At distances of 20 feet (6 m) to 40 feet (12 m), target area should be center mass
- This round is most effective against area targets with a maximum effective range of 15 yards
- Training needs to reinforce that head shots are NOT ACCEPTABLE
- Round is most effective against area targets (multiple personnel)
- Stock Sock: a yellow "stock sock" will be used whenever the modified 12-gauge Remington 870 shotgun is loaded with nonlethal munitions. This removable nylon sleeve is used to identify the weapon as a nonlethal delivery system.

- All specifications are average and are subject to change without notice
- 2552SPEC

Figure B-11. Cartridge, 12-Gauge Rubber Sting Ball™

Cartridge, 12-Gauge Sting Ball™ High Velocity

Using Services: USCG
Status: Fielded
DODIC: N/A Model No. 2553
NSN: Commercial Procurement
Acquisitions Activity: Commandant (G-OCU)
Further Information: Commandant (G-OCU) (202) 267-1522

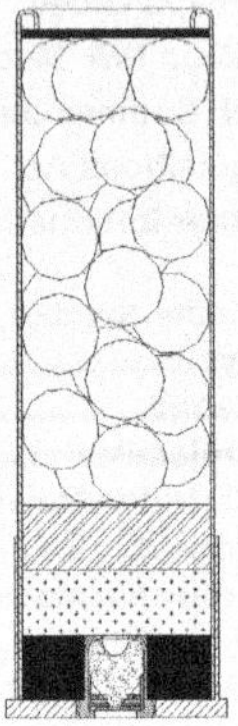

Model Number .... 31-caliber sting-ball
Type .... Multiple projectile rubber pellets
Range .... 10 yards (9 m)
Projectiles .... 18 (average)
Cartridge Material................. Plastic
Projectile Weight................... .35 grams each
Projectile Material.................. .31-caliber rubber pellets
Overall Height........................ 2.4 to 2.5 inches (61 – 63 mm)
Overall Weight .... 18 grams
Compatibility.......................... 12-gauge shotguns
Velocity.................................. 900 feet/second

USCG

- Fatal injuries are possible at employment distances of less than 5 yards.
- At distances of 20 feet (6 m) to 40 feet (12 m), target area should be center mass
- Maximum effective range is 15 yards
- Training needs to reinforce that head shots are NOT ACCEPTABLE
- Round is most effective against area targets (multiple personnel)
- Stock Sock: a yellow "stock sock" will be used whenever the modified 12-gauge Remington 870 shotgun is loaded with nonlethal munitions. This removable nylon sleeve is used to identify the weapon as a nonlethal delivery system.

- All specifications are average and are subject to change without notice.
- 2553SPEC

Figure B-12. Cartridge, 12-Gauge Sting Ball™ High Velocity

M1006, 40-mm Nonlethal Cartridge (Sponge Grenade)

Using Services: U.S. Army

Status: Materiel Released April 2000

DODIC: BA06

NSN: 1310-01-452-1190

Acquisition Activity: Acquisition Center, Picatinny Arsenal, NJ

Further Information:

Commercial: Picatinny Arsenal NL Systems Integrator: (973) 724-6948
Product Director: (973) 724-6283

Color:

- Projectile .... Black, olive green (foam rubber)
- Cartridge Case .... Opaque

Diameter .... 40 millimeter

Length .... 5.272 inches (13.4 cm) (maximum)

Total Weight .... 68 grams (0.15 lbs)

Propellant Weight .... 2 grains

Muzzle Velocity .... 265 + 20 feet/second (81 + 6 m/sec)

Acceptance Accuracy .... >85 % accuracy at 30 meters (E-silhouette) and >60 % accuracy at 50 meters (E-silhouette) when fired from an M16A2/M203 or a M4/M203

Engagement Range .... 10 meters to 50 meters

Projectile Specification:

- Shape .... 40-mm bullet shaped foam rubber
- Material .... Foam rubber
- Weight .... 28.5 grams (.06 lbs)

- Fatal injuries are possible at employment distances of less than 32.8 feet (10 m)
- At distances of 10 meters to 50 meters, target area should be center mass
- Training needs to reinforce that head shots are NOT ACCEPTABLE
- Do not skip fire this round! Round becomes unpredictable upon striking the ground
- Round is most effective against individually selected targets (point round)

Figure B-13. M1006, 40-mm Nonlethal Cartridge (Sponge Grenade)

M1012, 12-Gauge Shotgun: Nonlethal, Point Control Cartridge

Using Services: U.S. Army

Status: Type Classified, June 2000

DODAC: AA51

NSN: 1305-01-470-2405

Acquisition Activity: Acquisition Center, Picatinny Arsenal, NJ

Further Information:

Commercial: Picatinny Arsenal NL Systems Integrator: (973) 724-6948
Product Director: (973) 724-6086

Color:
- Cartridge Case: .... Clear

Diameter .... 12 gauge
Length .... 2.45 inches (62.2 mm)
Total Weight .... 500 grains
Propellant Weight .... 6.6 grains smokeless (Red Dot)
Muzzle Velocity .... 500 feet/second (152.4 m/sec)
Acceptance Accuracy .... 80 % accuracy at 10 and 20 meters (E-silhouette) fired out of Mossberg 500, Mossberg 590, and Winchester 1200
Engagement range .... 10 to 20 meters

Projectile Specification:
- Shape .... Fin stabilized 'bomblet' shape
- Material .... Rubber
- Weight .... 0.2 ounces (5.8 gm)
- Hardness .... 75 Durometer "A" Scale
  - Fatal injuries are possible at employment distances of less than 33 feet (10 m)
  - At distances of 10 to 20 meters, target area should be center mass
  - Training needs to reinforce that head shots are NOT ACCEPTABLE
  - Round is most effective against individually selected targets (point round)

Figure B-14. M1012, 12-Gauge Shotgun: Nonlethal, Point Control Cartridge

M1013, 12-Gauge Shotgun: Nonlethal, Crowd Dispersal Cartridge

Using Services: U.S. Army

Status: Type Classified June 2000

DODAC: AA52

NSN: 1305-01-470-2139

Acquisition Activity: Acquisition Center, Picatinny Arsenal, NJ

Further Information:

Commercial: Picatinny Arsenal NL Systems Integrator: (973) 724-6948
Product Director :(973) 724-6086

Color:
- Projectile .... Black rubber
- Cartridge Case .... Clear

Diameter .... 12 gauge
Projectile .... 18 rubber balls
Length .... 2.37 inches (60.2 mm)
Total Weight.......................... 500 grains (32.4 gm)
Propellant Weight................. 14 grains smokeless (Red Dot)
Muzzle Velocity..................... 900 feet/second (274 m/sec)
Acceptance Accuracy.......... 80 % accuracy for a 2-by-3-meter target at 10 and 20 meters fired out of Mossberg 500, Mossberg 590, and Winchester 1200
Engagement range............... 10 to 20 meters

Projectile Specification:
- Shape............................... 32-gauge (0.32 in) ball
- Material............................ PVC rubber compound
- Weight.............................. 0.01 ounces (0.25 gm)
- Hardness .... 75 Durometer "A" Scale

- Fatal injuries are possible at employment distances of less than 32.8 feet (10 m)
- At distances of 10 to 20 meters, target area should be center mass
- Training needs to reinforce that head shots are NOT ACCEPTABLE

Figure B-15. M1013, 12-Gauge Shotgun: Nonlethal, Crowd Dispersal Cartridge

M1029, 40-mm Nonlethal Crowd Dispersal Grenade

Using Services: U.S. Army

Status: Type Classified May 2001

DODIC: BA13

NSN: 1310-01-475-0628

Acquisition Activity: Acquisition Center, Picatinny Arsenal, NJ

Further Information:

Commercial: Picatinny Arsenal NL Systems Integrator: (973) 724-6948
Product Director: (973) 724-6283

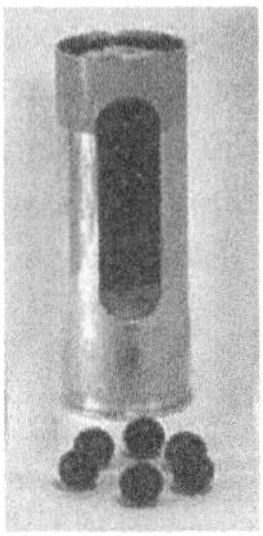

Color:

- Projectile .... Black rubber
- Cartridge Case .... Aluminum

Diameter .... 40 millimeter

Projectile .... 48 rubber balls

Length .... 4.8 inches (12.2 cm)

Total Weight.......................... 202.22 grams (7.0 oz)

Propellant Weight................. 8.5 grains smokeless

Muzzle Velocity..................... 450 feet/second

Acceptance Accuracy.......... > 90 % hit accuracy 4 out of 5 (shoulder to shoulder) E-silhouettes target at 30 meters and when fired from an M16A2/M203 or a M4/M203

Projectile Specification:

- Shape .... 48-gauge ball (.48 in diameter)
- Material.............................. Rubber
- Weight .... 1.3 grams (.046 oz)
- Hardness........................... Durometer 60, scale "A"

- Fatal injuries are possible at employment distances of less than 32.8 feet (10 m)
- At distances of 10 to 30 meters, target area should be center mass of group of individuals
- Round is most effective when used to disperse groups of individuals

Figure B-16. M1029, 40-mm Nonlethal Crowd Dispersal Grenade

L96A1*, Grenade, Discharger: Antiriot, Irritant, CS

Using Services: U.S. Army, USMC

Status: Materiel Release, 4 January 2002

DODIC: FZ14

NSN: 1330-01-459-4018

Acquisition Activity: Acquisition Center, Picatinny Arsenal, NJ

Further Information:

Commercial: Picatinny Arsenal NL Systems Integrator: (973) 724-6948
Project Officer: (410) 436-2139

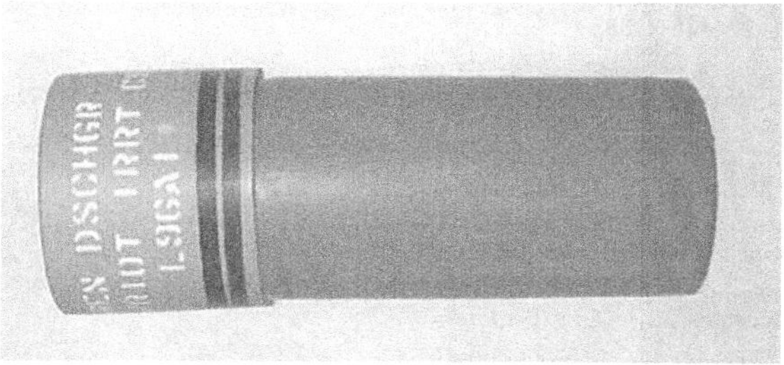

Color....................................Light gray with red and brown bands
Diameter...............................66 millimeters
Projectile..............................23 canisters filled with CS compound
Length..................................185 millimeters (7.28 in)
Total Weight..........................568 grams (1.25 lbs)
Propellant Weight................. 3.8 grams (.008 lbs)
Muzzle Velocity..................... 35.8 meters/second (117.5 ft/s)
Acceptance Accuracy..........65 to 95 meters
Deployment Range...............65 to 95 meters
Projectile Specification:

NOTE: It is not accurate to refer to the submunitions of the L96A1 grenade as projectiles. The antiriot material (CS) is released from the submunitions as they burn on the ground.

Shape..............................Cylindrical
Material...........................Aluminum
Weight.............................11.3 grams (.025 lbs)

* The L97A1 Grenade is the training grenade for the L96A1 Grenade.

Figure B-17. L96A1, Grenade, Discharger: Antiriot, Irritant, CS

L97A1, Grenade, Discharger: Antiriot, Practice

Using Services: U.S. Army, USMC

Status: Material Released, 4 January 02

DODIC: FZ15

NSN: 1330-01-459-4032

Acquisition Activity: Acquisition Center, Picatinny Arsenal, NJ

Further Information:

Commercial: Picatinny Arsenal NL Systems Integrator: (973) 724-6948 Project Officer: (410) 436-2139)

Color .... Blue with brown and green bands
Diameter .... 66 millimeters
Projectile .... 23 canisters filled with Cinnamic Acid (CA)
Length .... 185 millimeters (7.28 in)
Total Weight .... 568 grams (1.25 lbs)
Propellant Weight .... 3.8 grams (.008 lbs)
Muzzle Velocity .... 35.8 meters/second (117.5 ft/s)
Acceptance Accuracy .... 65 to 95 meters
Deployment Range .... 65 to 95 meters
Projectile Specification:

NOTE: It is not accurate to refer to the submunitions of the L97A1 grenade as projectiles. The antiriot, practice material (CA) is released from the submunitions as they burn on the ground.

Shape .... Cylindrical
Material .... Aluminum
Weight .... 11.3 gram (.025 lbs)

- CA smoke can cause irritation to the eyes, mucous membranes, and skin. Personnel entering a CA cloud must wear appropriate face protection such as an M17- or M40-series protective mask.
- The burning grenade canisters could start a fire in dry grasses. Keep fire extinguisher available.
- During firing training, personnel within 16 meters must wear single hearing protection. This includes personnel within vehicle.
- Do not fire grenades when personnel or equipment are within 150 meters covering a 90-degree arc around a firing discharger.
- Do not fire grenades at the 50- or 75-meter position on the M315 adjustable aiming bracket. Fire only at the 100-meter setting.

Figure B-18. L97A1, Grenade, Discharger: Antiriot, Practice

Grenade, Launcher: Nonlethal, Distraction, M98

Using Services: U.S. Army, USMC

Status: Type Classified, 17 October 2001

DODIC: FZ16

NSN: 1330-01-484-7773

Acquisition Activity: Acquisition Center, Picatinny Arsenal, NJ

Further Information: Commercial: Picatinny Arsenal
NL Systems Integrator:
(973) 724-6948
Project Officer: (410) 436-2139

Color .... Green and black with brown band
Diameter .... 66 millimeters
Projectile .... 3 bursting submunitions, with pyrotechnic charges for audio and visual stimuli
Length .... 25.25 centimeters (9.94 in)
Total Weight........................... 725.7 grams (1.6 lbs)
Propellant Weight................. 5.0 grams (.011 lbs)
Muzzle Velocity...................... 48.9 meters/seconds (160.4 ft/s)
Acceptance Accuracy.......... 100 to 120 meters
Deployment Radius.............. 10 to 15 meters
Projectile Specification:

NOTE: It is not accurate to refer to the submunitons of the M98 grenade as projectiles. These submunitions are ground bursting and the payload of inert ballast is just for flight stabilization and to deliver the audio/visual stimuli to the required range.

Filler .... Inert Ballast

- There is a remote chance of starting a fire. Keep fire extinguisher available.
- Personnel within 17 meters of a firing discharger must wear single hearing protection. This includes personnel within vehicle.
- Avoid looking directly at a bursting grenade; it could result in temporary loss of vision, lasting 2 to 5 minutes.
- Personnel within 35 meters of an operating launcher or 0.5 meters of a bursting canister should wear safety or ballistic-type eye protection, military clothing with long sleeves and elevated shirt collar, and helmet headgear.
- Do not handle or load M98s within 50 meters of a frequency transmitter emitting at 175 MHz, 448 MHz, 503 MHz and 1430 MHz.
- Do Not fire grenades when personnel or equipment are within 160 meters covering a 90-degree arc around a firing discharger.

Figure B-19. Grenade, Launcher: Nonlethal, Distraction, M98

M99, Grenade, Launcher: Nonlethal, Blunt Trauma

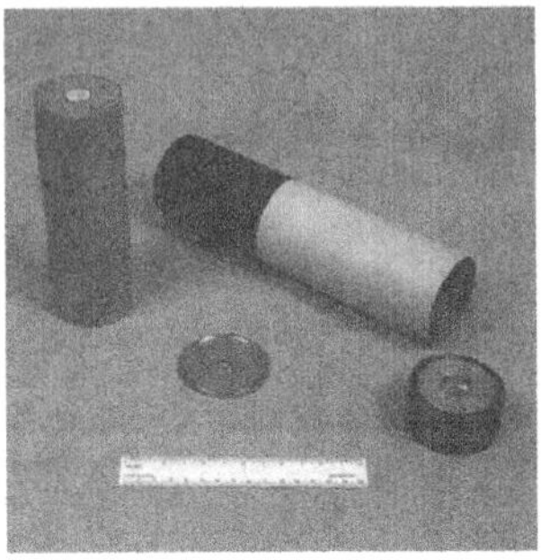

Using Services: U.S. Army, USMC

Status: Type Classified—17 October 2001

DODIC: FZ17

NSN: 1330-01-484-7775

Acquisition Activity: Acquisition Center, Picatinny Arsenal, NJ

Further Information:

Commercial: Picatinny Arsenal NL Systems Integrator: (973) 724-6948
Project Officer: (410-436-2139)

Color ....Green and black with brown band
Diameter ....66 millimeters
Projectile ....3 bursting submunitions, each with 140 balls
Length ....25.25 centimeters (9.94 in)
Total Weight..........................725.7 grams (1.6 lbs)
Propellant Weight.................5.0 grams (.011 lbs)
Muzzle Velocity.....................48.9 meters/second (160.4 ft/s)
Acceptance Accuracy..........100 to 120 meters
Deployment Radius..............5 to 10 meters
Projectile Specification:

NOTE: It is not accurate to refer to the submunitions of the M99 grenade as projectiles. These submunitions are ground bursting and the payload of inert ballast is just for flight stabilization and to deliver the audio/visual stimuli to the required range.

Shape...............................32-caliber (5/16 in) ball
Material ....PVC
Weight..............................0.40 grams (.014 oz)
Hardness ....90 Durometer

- There is a remote chance of starting a fire. Keep fire extinguisher available.
- Personnel within 17 meters of a firing discharger must wear single hearing protection. This includes personnel within vehicle.
- Avoid looking directly at a bursting grenade; it could result in temporary loss of vision, lasting 2 to 5 minutes
- Personnel within 35 meters of an operating launcher or 0.5 meters of a bursting canister should wear safety or ballistic-type eye protection, military clothing with long sleeves and elevated shirt collar, and helmet headgear.
- Do not handle or load M98s within 50 meters of a frequency transmitter emitting at 175 MHz, 448 MHz, 503 MHz and 1430 MHz.
- DO NOT fire grenades when personnel or equipment are within 160 meters covering a 90-degree arc around a firing discharger.
- To minimize risk of eye injury, use the M99 primarily against crowds of adults.

Figure B-20. M99, Grenade, Launcher: Nonlethal, Blunt Trauma

Appendix C

Interservice Nonlethal Individual Weapons Instructor Course

1. Introduction

This appendix provides an overview of the Interservice Nonlethal Individual Weapons Instructor Course (INIWIC). The INIWIC is a "train-the-trainer" course taught in 15 subcourses and divided into four sections. Section I provides INIWIC descriptive data that includes the purpose, scope, peacetime and mobilization training requirements, student prerequisites, and equipment/ammunition requirements necessary to successfully complete the course. Section II lists the total training days and hours required per academic subcourse. Section III gives a detailed description of each subcourse and terminal learning objectives that each student must meet to complete the course. Section IV is a listing of 15 annexes that comprise the INIWIC Program of Instruction (POI). Access the entire INIWIC, with history, updates, and lesson plans at the following Internet address: http://www.wood.army.mil/mcdet/niwic/niwic.htm.

2. Course Descriptive Data

a. Course Title: Interservice Nonlethal Individual Weapons Instructor Course

b. Location: Marine Corps Detachment, Fort Leonard Wood, Missouri

c. Marine Corps service school code: A16H5A3

d. Other service course number: USAF—L5AZA3P071019

e. Military assistance program articles and service list number: P166810

f. Purpose: To certify selected DOD/DOT service persons as nonlethal individual weapons instructors.

g. Scope: This is a train-the-trainer course designed for E-4s and above. The objective of this course is to prepare students to train individuals and groups in the utilization and employment of NLW and to serve as the unit NLW subject matter expert. This course is designed for all occupational specialties.

h. Length: 10 training days

i. Curriculum Breakdown: 100 total hours.

(1) 2.00 Demonstration.

(2) 24.00 Demonstration/practical exercise.

(3) 22.50 Lecture.

(4) 5.00 Lecture/demonstration.

(5) 4.00 Live-fire exercise.

(6) 26.50 Practical application.

(7) 6.00 Performance exam.

(8) 4.00 Written exam.

(9) 6.00 Administrative time.

j. Maximum Class Capacity: 30.

k. Optimum Class Capacity: 25.

l. Minimum Class Capacity: 15.

m. Class Frequency: 12.

n. Prerequisites: Students must be selected by commanders to be nonlethal weapons instructors and be an E-4 or above.

o. MOS Received: None.

p. Quota Control: Formal Schools Branch, Training and Education Command, Quantico, Virginia.

q. Funding: Service or command funded.

r. Reporting Instructions: All students report with orders to the Fort Leonard Wood, Missouri, Billeting Office, Bldg # 470 no later than 2400 on the day prior to class convening. Government quarters are available; however, messing is not.

s. Instructor Staffing: Two USMC instructors and two U.S. Army instructors.

t. School Overhead: Course OIC and chief instructor

Training Support: Training support is provided by the USMC for all common training support items. Service-specific training items are provided by the respective service.

Table C-1. Required Equipment For INIWIC

Item	*Quantity*
Personnel Protectors:	
Non-Ballistic Riot Face Shield	40
Non-Ballistic Full Length Riot Shield	40
Non-Ballistic Shin Guards	40
Ballistic Riot Face Shield	20
Ballistic Full Length Riot Shield	20
Ballistic Shin Guards	20
Kevlar Helmets	40
Flak Jacket/Body Armor	40
Personnel Effectors:	
Disposable Handcuffs (Flex Cuffs)	300
12-Gauge Area (rubber ball)	1400
12-Gauge Beanbag	1400
12-Gauge Fin Stabilized	1400
12-Gauge Flash-bang	1400
12-Gauge OO Buck	xxxx
12-Gauge 7 1/2 shot	xxxx
12-Gauge Launching Cup Cartridge	1400
40-mm Area (rubber ball)	1400
40-mm Foam Baton (3 foam sub-munitions)	1400
40-mm Sponge Grenade	1400
MK141 Diversionary Grenade	600

Table C-1. Required Equipment For INIWIC

Item	*Quantity*
M84 Reloadable Diversionary Grenade	600
Sting Ball Hand Thrown or Shot launched Grenade	600
Mossberg 500A2 Shotgun Launch Cup	15
M203 Grenade Launcher	15
Mossberg 500 3" Shotgun	15
36" Riot Oak Baton w/Belt Ring	40
Monadnock® 24-36" Expandable Baton w/holster	40
M36 RCA Dispenser	40
Mk-4 Personal OC Dispenser	40
Mk-9 Team OC Dispenser	40
Mk-46 High Capacity High Output OC Dispenser	20
MK-46 Hardware Kit (Refill Kit)	10
M33A1 Squad Riot Control Agent Dispenser	10
Modular Crowd Control Munitions Ground Emplaced (MCCM-GE)	20
66-mm Vehicle Launched Nonlethal Grenade (VLNLG)	3
66-mm NL munitions	30
Mission Enhancers:	
Portable Bullhorn	5
40-mm Carrying Pouch	40
Diversionary Rubber Ball Grenade Pouch	40
12-Gauge Butt Cuff carriers	40
12-Gauge Utility 25-Round Pouch	40
Mk-4 Individual Carrying Pouch	40
Portable Public Address System w/Microphone	3
12-Volt Hand-Held Spotlight	5
Night Hunter High Intensity Light System w/Kit	5
12-Gauge High Intensity Light Set	5
Caltrops	5
Road Side Spike Strip	5
Portable Vehicle Arrest Barrier (PVAB)	3
Training Devices:	
Impact Training Suit	6
Striking Bags	15
Training Batons (24 in)	40
Training Batons (36 in)	40
Practice Grenade Body	40
Practice Grenade Fuse	540
• OPFOR Training Munitions (foreign objects)	xxxx
• Target Requirements (full or partial torso)	40
MK-4 Personal OC Dispenser (training)	300
MK-9 Team OC Dispenser (training)	300
Training Disposable Handcuffs (flex cuffs)	100
12-Gauge Dummy Rounds	100

Table C-2. INIWIC Class V Requirements

Type	*#Per Student*	*# Demo Rounds*	*Total Per Class*	*Total Per Year*
12-Gauge Area (rubber ball)	5	5	155	1400
12-Gauge Beanbag	5	5	155	1400
12-Gauge Fin Stabilized	5	5	155	1400
40-mm Area (rubber ball)	5	5	155	1400
40-mm Foam Baton	5	5	155	1400
Sting Ball Hand Thrown or Shotgun Launched Grenades	2	2	62	600
MK141 Diversion Devices	2	2	62	600

3. Training Days Summary

This course consists of 10 training days. The curriculum is broken down into three categories.

Table C-3. INIWIC Curriculum

NLW Tactics	*Hours*
Force Continuum/ROE	2.00
Crowd Dynamics/Control/Formations	14.00
Communication Skills	2.00
MOOTW	2.00
Tactical Considerations	2.00
	22.00
NLW Systems	***Hours***
Oleoresin Capsicum	4.00
Empty Hand Control Techniques	19.00
Riot Baton Techniques	8.00
Nonlethal Munitions and Equipment	8.00
Portable Vehicle Barrier	5.00
66-mm Grenade Launcher	2.00
Modular Crowd Control Munition	2.00
	48.00
Nonlethal Weapons Instructor	***Hours***
Techniques of Military Instruction	10.00
NLW Scenario Exercise	10.00
	20.00
Subtotal =	**90.00**
Examinations =	***4.00***
Administration =	***6.00***
Total Hours =	***100.00***

4. Course Body

The course body consists of 15 subcourses. A quick synopsis of each subcourse is included below.

a. **Instructor Development**. This subcourse teaches and reinforces basic military instructor methodology. It identifies characteristics of good instructors and four ways they can improve their teaching ability. Additionally, it illustrates rules of conducting military instruction as well as the steps in planning a lesson and teaching classes.

b. **Force Continuum.** This subcourse introduces the five levels of resistance as well as identifying the levels of the force continuum. It illustrates influencing the use of force decisions and the definitions relating to deadly force. Lastly, it identifies the three preconditions for deadly force and seven authorized occasions for use of deadly force.

c. **Crowd Dynamics/Crowd Control**. This subcourse outlines the differences between crowds, mobs, and riots, and teaches the student basic crowd control techniques to apply in various situations. Upon completion, the student will be able to instruct others in both the behavioral aspects of a crowd during a civil disturbance and also various techniques of controlling a civil disturbance. It will identify factors that affect both individual and crowd behavior and it identifies all crowd control equipment needed, or essential, for mission accomplishment.

d. **Empty Hand Control Techniques**. This subcourse identifies safety factors to consider when instructing empty hand control techniques by illustrating the four Ms (mind, mass, movement, muscle) of self-defense, psychomotor skills, and phases of psychomotor skills. It illustrates levels of training and identifies common pressure points of the human body. It identifies hinges, appropriate angles of approach, and recommended patterns of movement. Instructors will learn blocking techniques, striking skills, control holds, escort positions, and restraint devices.

e. **Baton Techniques**. This subcourse identifies riot baton nomenclature and safety standards when conducting riot baton training. It identifies methods of teaching psychomotor skills and the phases of teaching psychomotor skills. It highlights baton-striking techniques, blocking techniques, and retention techniques.

f. **Military Operations Other Than War (MOOTW)**. This subcourse defines operations other than war and the difference between MOOTW and declarations of war. It helps identify the primacy of political objectives and the strategic aspects of MOOTW. It illustrates the range of operations and the principles of operations, as well as the types of military operations other than war.

g. **Rules of Engagement (ROE)**. This subcourse defines ROE and specific terms that apply to these rules. It also defines the purpose of ROE and provides the student an understanding of policies underlying the need for ROE. It identifies three sources for ROE and the two parts of ROE that will not change. It explains how ROE relate to using and applying NLW.

h. **Oleoresin Capsicum (OC) Training**. This subcourse explains how the three levels of OC are created, as well as the nomenclatures for all aerosol canisters. It highlights the different spray patterns, methods of delivery, and cautions concerning the traumatic hydraulic needle effect. It explains employment considerations and the three levels of OC contamination. Additionally, it explains the physical and mental effects of OC and trains students on different grips, draw methods, and proper stances when employing OC. It explains procedures for decontamination, recovery, and first aid.

i. **Types of Formations**. This subcourse explains the elements and individual responsibilities within a riot control formation. Additionally, it trains proper grip, balance, holding, and retention of the riot control shield. It highlights leadership requirements by teaching different commands and cadence for conducting riot control formations. It illustrates the different types of formations and the support elements of a riot control operation

j. **Tactical Considerations**. This subcourse addresses the preparations necessary for crowd control operations and the graduated response of a crowd control force. It identifies support and information required from host nation police and civil authorities. It describes the employment measures of military forces in crowd control and MOOTW environments. It highlights the actions, optional measures, and guidance for using NLW in crowd control and MOOTW environments.

k. **Nonlethal Munitions and Employment**. This subcourse identifies the types of nonlethal munitions and their intended purpose. It illustrates how nonlethal munitions fall within established "use of force" continuums. It identifies the types, characteristics, and ranges of all currently employed nonlethal munitions.

l. **Communications Skills**. This subcourse identifies the areas of nonverbal communication employed to determine an individual's state of mind. This, along with an understanding of the importance of personal space, gives the student the tools to resolve conflict with the amount of force as applied in the force continuum.

m. **Portable Vehicle Arrest Barrier**. This subcourse teaches the employment and maintenance of a force protection asset assigned to the nonlethal capability set. The PVAB is designed to stop a 7,500-pound vehicle going 45 mph. The PVAB protects gates and check points.

n. **Modular Crowd Control Munition**. This subcourse teaches the employment of the MCCM; the nonlethal version of the claymore mine. The MCCM is a force protection asset designed to help protect static defensive positions or check points.

o. **M-315 66-mm Grenade Launcher**. This subcourse teaches the installation, employment, and maintenance of the M-315. The M-315 increases the current maximum range of nonlethal sting ball munitions from approximately 50 meters to 100 meters. Additionally, this weapons platform can deploy flash bangs, smoke, and CS munitions.

Appendix D

Sample: Unit Training Schedules/Firing Tables

Nonlethal Individual Weapons Training Course Training Schedule: Unit Training Schedule Number 1.

Training Day #1					
Time	***Subject***	***Location***	***Uniform***	***Instr***	***Ref/Notes***
0800	Administration Time	Classroom	U1		
0830	Force Continuum	Classroom	U1		
1000	ROE	Classroom	U1		
1200	Chow	Mess Hall	U1		
1300	MOOTW	Classroom	U1		
1500	Communication Skills	Classroom	U1		
1700	Secure/COB				

Administration:

Uniforms:

U1: BDU/utility uniform
U2: BDU/utility uniform with LBE/harness and canteens
U3: BDU/utility uniform with LBE/harness, canteens, flak jacket, and helmet
U4: BDU/utility uniform with LBE/harness, canteens, flak jacket, helmet, and assigned weapon

Transportation:
Weapons:
Ammunition:
Required Equipment:
Screen
Video projector with central processing unit
Televisions
VCR
Miscellaneous:
All personnel are required to have a notebook, pencil, and pen.
Handouts prepared and passed out.

Training Day #2					
Time	***Subject***	***Location***	***Uniform***	***Instr***	***Ref/Notes***
0800	Nonlethal Equipment Brief	Classroom	U1		
0900	Pressure Point Control Techniques/ Open Hand Control Techniques	Classroom	U1		
1000	Pressure Point Control Techniques/ Open Hand Control Techniques (Practical Application)	PT Field	U2		
1200	Chow				
1300	Pressure Point Control Techniques/ Open Hand Control Techniques (Practical Application)	PT Field	U2		
1700	Debrief/Secure/COB				

Administration:

Uniforms:
U1: BDU/utility uniform
U2: BDU/utility uniform with LBE/harness and canteens
U3: BDU/utility uniform with LBE/harness, canteens, flak jacket, and helmet
U4: BDU/utility uniform with LBE/harness, canteens, flak jacket, helmet, and assigned weapon

Transportation:

Weapons:

Ammunition:

Required Equipment:
Screen
Video projector with central processing unit
Flexicuffs (one set of cuffs per two students)
Riot Training Bags (one bag per two students)

Miscellaneous:
All personnel are required to have a notebook, pencil, and pen.
Handouts prepared and passed out.
Medical personnel on station.

Training Day #3					
Time	***Subject***	***Location***	***Uniform***	***Instr***	***Ref/Notes***
0800	Pressure Point Control	Classroom	U1		
	Techniques/Open Hand Control Techniques	Classroom	U1		
0900	Pressure Point Control Techniques/Open Hand Control Techniques (Practical Application)	PT Field	U1		
1200	Chow				
1300	Pressure Point Control Techniques/Open Hand Control Techniques (Practical Application)	PT Field	U2		
1700	Debrief/Secure/COB				

Administration:

Uniforms:
U1: BDU/utility uniform
U2: BDU/utility uniform with LBE/harness and canteens
U3: BDU/utility uniform with LBE/harness, canteens, flak jacket, and helmet
U4: BDU/utility uniform with LBE/harness, canteens, flak jacket, helmet, and assigned weapon

Transportation:

Weapons:

Ammunition:

Required Equipment:
Screen
Video projector with central processing unit
Flex cuffs (one set of cuffs per 2 students)
Riot Training Bags (one bag per 2 students)

Miscellaneous:
All personnel are required to have a notebook, pencil, and pen.
Handouts prepared and passed out.
Medical personnel on station.

Training Day #4

Time	*Subject*	*Location*	*Uniform*	*Instr*	*Ref/Notes*
0800	Impact Weapons	Classroom	U1		
0900	Impact Weapons (Practical Application)	PT Field	U2		
1200	Chow				
1300	Impact Weapons (Practical Application)	PT Field	U2		
1700	Debrief/Secure/COB				

Administration:

Uniforms:
U1: BDU/utility uniform
U2: BDU/utility uniform with LBE/harness and canteens
U3: BDU/utility uniform with LBE/harness, canteens, flak jacket, and helmet
U4: BDU/utility uniform with LBE/harness, canteens, flak jacket, helmet, and assigned weapon

Transportation:

Weapons:
Baton (one baton per 2 students)
Training baton (one baton per 2 students)

Ammunition:

Required Equipment:
Screen
Video projector with central processing unit
Flex cuffs (one set of cuffs per 2 students)
Riot training bags (one bag per 2 students)
Three (3) riot training suits

Miscellaneous:
All personnel are required to have a notebook, pencil, and pen.
Handouts prepared and passed out.
Medical personnel on station.

Training Day #5					
Time	***Subject***	***Location***	***Uniform***	***Instr***	***Ref/Notes***
0800	Impact Weapons (Practical Application)	PT Field	U2		
1000	Crowd Dynamics/Crowd Control	Classroom	U1		
1130	Chow				
1230	Crowd Dynamics/Crowd Control	Classroom	U1		
1330	Riot Control Formations	Classroom	U1		
1430	Oleoresin Capsicum (OC) Aerosol Training	Classroom	U1		
1630	Debrief/Secure/COB				

Administration:

Uniforms:
U1: BDU/utility uniform
U2: BDU/utility uniform with LBE/harness and canteens
U3: BDU/utility uniform with LBE/harness, canteens, flak jacket, and helmet
U4: BDU/utility uniform with LBE/harness, canteens, flak jacket, helmet, and assigned weapon

Transportation:

Weapons:
Training baton (one baton per 2 students)
Baton (one baton per 2 students)

Ammunition:
Inert OC for classroom instruction

Required Equipment:
Riot training bags (one bag per 2 students)
Screen
Video projector with central processing unit
Three (3) riot training suits

Miscellaneous:
All personnel are required to have a notebook, pencil, and pen.
Handouts prepared and passed out.

Training Day #6					
Time	***Subject***	***Location***	***Uniform***	***Instr***	***Ref/Notes***
0800	OC Aerosol Training (Practical Application)	PT Field	U2		
1200	Chow				
1300	OC Aerosol Training (Practical Application)	PT Field	U2		
TBD	Debrief/Secure/COB				

Administration:

Uniforms:
U1: BDU/utility uniform
U2: BDU/utility uniform with LBE/harness and canteens
U3: BDU/utility uniform with LBE/harness, canteens, flak jacket, and helmet
U4: BDU/utility uniform with LBE/harness, canteens, flak jacket, helmet, and assigned weapon

Transportation:

Weapons:

Ammunition:
MK 4.inert OC (one per student)
MK 4 individual can OC (2 cans)

Required Equipment:
Five (5) large trash cans with liners
Two (2) water hoses with running water

Miscellaneous:
Medical personnel on station.

Training Day #7

Time	Subject	Location	Uniform	Instr	Ref/Notes
0730	Movement to Range		U4		
0800	Nonlethal Munitions and Employment	Range Classroom	U1		
1030	Chow	Range			
1100	Safety Brief/Range Brief	Range	U4		
1130	Nonlethal Munitions (Live Fire)	Range	U4		
TBD	Evening Chow				
TBD	Safety Brief/Range Brief	Range	U4		
1800	Night Exercise Sun Set +30)	Range	U4		
TBD	Debrief/Clean Up	Range	U1		
TBD	Movement to Rear				
TBD	Secure/ COB				

Administration:

Uniforms:

U1: BDU/utility uniform
U2: BDU/utility uniform with LBE/harness and canteens
U3: BDU/utility uniform with LBE/harness, canteens, flak jacket, and helmet
U4: BDU/utility uniform with LBE/harness, canteens, flak jacket, helmet, and assigned weapon

Transportation: Yes

Weapons:

Assigned Weapons

Ammunition:

12-gauge rubber ball
12-gauge fin stabilized
12-gauge bean bag
12-gauge launching cup cartridge
40-mm foam baton
40-mm rubber ball
40-mm sponge grenade
Stingball grenades

Required Equipment:

Eight (8) 12-Gauge stinger launcher cup
Fifteen (15) targets
Eye and hearing protection
Fifteen (15) full-length riot shields
All students require munitions pouches, shin guards, face shields
Two (2) MRE per student
Video projector with central processing unit
Three (3) high-intensity light
Thirty (30) individual tactical lights

Miscellaneous:

Range safety officer.
Medical personnel on station.
Secondary means of communication (radios).
Conduct operational risk assessment.

Training Day #8					
Time	*Subject*	*Location*	*Uniform*	*Instr*	*Ref/Notes*
0700	Movement to Training Area #1		U2		
0730	Portable Vehicle Arresting Barrier (PVAB)	TA#1	U1		
0900	PVAB (Practical Application)	TA#1	U2		
1200	Chow				
1300	PVAB (Practical Application)	TA#1	U2		
1500	Debrief/Movement to Rear				
1530	Preventive Maintenance of PVAB				
1600	Secure/COB				

Administration:

Uniforms:
U1: BDU/utility uniform
U2: BDU/utility uniform with LBE/harness and canteens
U3: BDU/utility uniform with LBE/harness, canteens, flak jacket, and helmet
U4: BDU/utility uniform with LBE/harness, canteens, flak jacket, helmet, and assigned weapon

Transportation: Yes

Weapons:

Ammunition:

Required Equipment:
One (1) portable vehicle-arresting barrier (PVAB)
Video projector with central processing unit

Miscellaneous:
(1) MRE per student.

Training Day #9					
Time	***Subject***	***Location***	***Uniform***	***Instr***	***Ref/Notes***
0730	Movement to Range		U4		
0800	Modular Crowd Control Munition-Ground Emplaced (MCCM-GE)	Range Classroom	U1		
0900	Safety Brief/Range Brief	Range	U4		
0930	MCCM-GE (Practical Application	Range	U4		
1030	Chow				
1130	Vehicle Mounted Nonlethal Grenade Launcher (VMNLGL)	Range Classroom	U4		
1300	VMNGL	Range	U4		
1730	Debrief/Movement to Rear				
1800	Secure/COB				

Administration:

Uniforms:

U1: BDU/utility uniform
U2: BDU/utility uniform with LBE/harness and canteens
U3: BDU/utility uniform with LBE/harness, canteens, flak jacket, and helmet
U4: BDU/utility uniform with LBE/harness, canteens, flak jacket, helmet, and assigned weapon

Transportation: Yes

Weapons:

Vehicle Mounted Nonlethal Grenade Launcher (VMNLGL)

Ammunition:

(2) MCCM-GE, (2) 66-mm Riot Control Training Grenade (L97E1)

Required Equipment:

Eye and ear protection
Video projector with central processing unit

Miscellaneous:

Range safety officer.
Medical personnel on station.
Secondary means of communication.
Two (2) MRE per student.

Training Day #10

Time	*Subject*	*Location*	*Uniform*	*Instr*	*Ref/Notes*
0700	Movement Down Range				U4
0730	Tactical Considerations	MOUT Classroom		U1	
0930	Safety Brief/Range Brief	MOUT Facility		U4	
1000	Tactical Exercise* (Practical Application)	MOUT Facility		U4	
1200	Chow				
1300	Tactical Exercise (Practical Application)	MOUT Facility		U4	
1700	Debrief/Movement to Rear				
1730	Weapons Maintenance				
1800	Secure/COB				

Administration:

Uniforms:
U1: BDU/utility uniform
U2: BDU/utility uniform with LBE/harness and canteens
U3: BDU/utility uniform with LBE/harness, canteens, flak jacket, and helmet
U4: BDU/utility uniform with LBE/harness, canteens, flak jacket, helmet, and assigned weapon

Transportation: Yes

Weapons:
Assigned weapon
Inert OC individual (one per student)
Training batons (one per student)
Fifty (50) practice grenades

Ammunition:

Required Equipment:
Three (3) riot training suit
All students require munitions pouches, flex cuffs, shin guards, face shields
Video projector with central processing unit
Fifteen (15) full length riot shields

Miscellaneous:
Range safety officer.
Medical personnel present.
Secondary means of communication.
(2) MRE's per student.

UNITED STATES MARINE CORPS
Marine Corps Detachment
Fort Leonard Wood, MO 65473

1.	LESSON DESIGNATOR:	Interservice Nonlethal Individual Weapons Instructor Course (INWIC). 13A (INIWIC.13A)
2.	LESSON TITLE:	Scenario Practical Exercise
3.	SCHOOL/COURSE:	Nonlethal Individual Weapons Instructor Course
4.	STUDY ASSIGNMENT:	Review course curriculum
5.	ESTIMATED TIME:	8 hours
6.	INSTRUCTORS REQUIRED:	(1) primary, (1) alternate per ratio: 1:10
7.	INSTRUCTOR REFERENCES:	INIWIC POI
8.	DATE PREPARED:	September 2001
9.	PRESENTATION METHOD:	Practical Exercise
10.	TRAINING AIDS:	Warning Order and five paragraph operations order for exercise, rubber rifles/shotguns, batons, shields, Inert MK4 OC canister, bullhorn, radio
11.	STUDENT EQUIPMENT:	(1) 782 gear (1) Kevlar with face shield (1) Shin guards

5. Unit Training Schedule Number 2

Table D-1. NLW Training Course 5-Day Training Schedule

Day/date	*Time*	*Event*	*Location*	*Instructor*	*Notes*
Mon/07 Jan 02	0730-0745	Welcome Aboard	Bldg RR202	MSgt Dixon	
T-1	0745-0845	Force Continuum	SOTG	TBD	L
	0900-1015	MOAB	SOTG	TBD	L
	1030-1130	OC	SOTG	TBD	L
	1130-1300	Noon Meal	Mess Hall	Company GySgt	
	1300-1400	OC Practical Application	SOTG	TBD	PA
	1400-1700	OC Certification	SOTG	TBD	PA
Tues/08 Jan 02	0730-0830	ROE	SOTG	TBD	L
T-2	0845-0945	Crowd Control	SOTG	TBD	L
	1000-1130	MDTS	SOTG	TBD	L, PA
	1130-1300	Noon Meal	Mess Hall	Company GySgt	
	1300-1700	MDTS	SOTG	TBD	L, PA
Wed/09 Jan 02	0730-1130	Batons	SOTG	TBD	L, PA
T-3	1130-1300	Noon Meal	Mess Hall	Company GySgt	
	1300-1500	Batons	SOTG	TBD	L, PA
	1500-1700	Formations	SOTG	TBD	L
Thurs/10 Jan 02	0730-1130	Formations	SOTG	TBD	PA
T-4	1130-1300	Noon Meal	Mess Hall	Company GySgt	
	1300-1700	Nonlethal Munitions/shotguns/ M2O3/sustainment training	Square bay/M203	TBD	L, PA
Fri/11 Jan 02	0730-1130	Nonlethal Munitions/stinger ball grenades/shotgun/M2 03	Square Bay/M203	TBD	PA
T-5	1130-1300	Noon Meal	Mess Hall	Company GySgt	
	1300-1700	Nonlethal Munitions/stinger ball grenades/shotgun/M2 03	Nonlethal Range 1&2	TBD	PA

Notes:

L = Lecture
PA = Practical Application

MDTS = Manadnock Defensive Tactics System
MOAB = Management of Aggressive Behavior
SOTG = Special Operations Training Group (USMC training unit)

6. Sample Unit Training Schedule Number 3

Table D-2. NLW Training Course 10-Day Training Schedule

DAY/DATE	***TIME***	***EVENT***	***LOCATION***	***INSTRUCTOR***	***NOTES***
Mon/29 Oct 01	0730-0745	Welcome Aboard	Bldg RR-202	MSgt Dixon	
T-1	0745-0845	Force Continuum	PT Field	Sgt Whitworth	L
	0900-1015	MOAB	PT Field	GySgt German	L
	1030-1130	OC	PT Field	Sgt Hoskins	L
	1130-1300	Noon Meal	Mess Hall	Company Gunny	
	1300-1400	OC Practical Application	PT Field	Sgt Hoskins	PA
	1400-1700	OC Certification	PT Field	Sgt Hoskins	PA
Tues/30 Oct 01	0730-0830	ROE	PT Field	Cpl Fewox	L
T-2	0845-0945	Crowd Control	PT Field	SSgt Parrish	L
	1000-1130	MDTS	PT Field	Sgt Johnson	L, PA
	1130-1300	Noon Meal	Mess Hall	Company Gunny	
	1300-1700	MDTS	PT Field	Sgt Johnson	L, PA
Wed/31 Oct 01	0730-1130	Batons	PT Field	Sgt Johnson	L, PA
T-3	1130-1300	Noon Meal	Mess Hall	Company Gunny	
	1300-1500	Batons	PT Field	Sgt Hoskins	L, PA
	1500-1700	Formations	PT Field	GySgt German	L
Thur/1 Nov 01	0730-1130	Formations	PT Field	Sgt Whitworth	PA
T-4	1130-1300	Noon Meal	Mess Hall	Company Gunny	
	1300-1700	Nonlethal Munitions	Nonlethal Range 1&2	GySgt German	L, PA
Fri/2 Nov 01	0730-1130	Nonlethal Munitions	Nonlethal Range 1&2	SSgt Parrish	PA
T-5	1130-1300	Noon Meal	Mess Hall	Company Gunny	
	1300-1700	Nonlethal Munitions	Nonlethal Range 1&2	SSgt Parrish	PA
Mon/5 Nov 01	0730-1130	Nonlethal Grenades Formation Fire	Nonlethal Range 1&2	GySgt German	L, PA
T-6	1130-1300	Noon Meal	Mess Hall	Company Gunny	
	1130-1700	Formation Fire/Sustainment Training	Nonlethal Range 1&2	SSgt Blackler	PA
Tues/6 Nov 01	0730-0830	Introduction to MOOTW	PT Field	Maj Desy	L
T-7	0830-1130	Vehicle Checkpoints/Vehicle Searches	PT Field	Sgt Whitworth	L, PA
	1130-1300	Noon Meal	Mess Hall	Company Gunny	
	1300-1500	Vehicle	PT Field	SSgt Parrish	PA

Table D-2. NLW Training Course 10-Day Training Schedule

DAY/DATE	*TIME*	*EVENT*	*LOCATION*	*INSTRUCTOR*	*NOTES*
		Checkpoints/Vehicle Searches			
	1500-1700	Roadblocks/Personnel Searches	PT Field	SSgt Blackler	L, PA
Wed/7 Nov 01	0730-1130	Roadblock/Personnel Searches	PT Field	SSgt Blackler	PA
T-8	1130-1300	Noon Meal	Mess Hall	Company Gunny	
	1300	Operations Order	Nonlethal Classroom	MSgt Dixon	Company Staff
Thurs/8 Nov 01	0730-1000	Confirmation Brief	PT Field	Maj Desy	
T-9	1000-1200	Insert	Combat Town	SSgt Parrish	
	1200-0000	STX	Combat Town	SSgt Parrish	
Fri/9 Nov 01	0001-1200	STX	Combat Town	SSgt Parrish	
T-10	1200-1300	Cleanup/Debrief	Combat Town	Company Gunny	
	1300	Graduation	Combat Town	SOTG OIC	

Notes:

L = Lecture

PA = Practical Application

MOAB = Management of Aggressive Behavior

MDTS = Manadnock Defensive Tactics System

STX = Situational Training Exercise

Appendix E

NLW Capability Sets

This appendix provides the current NLW capability sets for the services. These sets continue to be upgraded and changed according to service needs and the availability of new technologies. The U.S. Army and USMC capability sets are standard for the service, while the USAF,'USN, and USCG are still developing and fielding their sets.

1. U.S. Army NLW Capability Set

Table E-1 U.S. Army Nonlethal Capability Set (Version FY 2001)

	Name	*Quantity*	*Unit Cost*	*Item Total*
GSAPMDK5X250AFHC	Non-Ballistic Riot Face Shields	200	$42.50	$8,500.00
GSAA TIPMBS9	Non-Ballistic Riot Body Shields Paulson	150	$88.50	$13,275.00
GSAATIUSMCSG	Non-Ballistic Riot Shin Guards	150	$60.00	$9,000.00
PT702MTUSA	Ballistic Face Shields	10	$246.00	$2,460.00
PT2237	Ballistic Body Shields	10	$1,052.00	$10,520.00
GSASF6SL	Sure Fire Field Light	10	$185.82	$1,858.20
MON2800 w/BCS-1	36" Hickory Riot Baton with Grommet	200	$15.50	$3,100.00
EADBH	Eagle Duty Baton Holder OD	200	$5.50	$1,100.00
GSAATIA2M	Portable Bullhorn	10	$82.00	$820.00
AAPB35W1	Ground Mount Bullhorn	8	$815.00	$6,520.00
AAPBWH10001	Megavox Wireless MIC	8	$0.00	$0.00
AASS450	Megavox Stand	8	$0.00	$0.00
GSASF6P	Individual High Intensity Light	15	$42.74	$641.10
GSAATIEATAVLFP6POD	Eagle Sure Fire 6P Holster	15	$8.75	$131.25
GSASFP60	Individual High Intensity Light Accessory Kit	15	$12.53	$187.95
XENNCSUSA	High Intensity Light	8	$2,795.00	$22,360.00
NIKFLEXCUFF	Disposable Forearm/Ankle Cuffs (Single)	1000	$0.70	$700.00
NIKCUTTER	Flex Cuff Cutters	2	$9.00	$18.00
GSAATIEAMK4	MK-4 Pouch OD	200	$6.75	$1,350.00
GSADEF5149	Inert Individual Training (OC) Dispenser	200	$6.07	$1,214.00
GSASF622	12-Gauge Shotgun High Intensity Light	50	$139.65	$6,982.50
GSAATIEAUPMS	12-Gauge Utility (25-Round) Pouch	50	$23.65	$1,182.50
GSAATISSSCQTR	12-Gauge Buttcuff Carrier (6-Round)	50	$10.75	$537.50
GSAATISASDD60D	Diversionary/Sting Ball Grenade Pouch	25	$47.25	$1,181.25
GSAATIEA40MM6	40-mm Carrying Pouch	100	$13.65	$$1,365.00
GSAATIRSC3	Caltrop	250	$10.50	$2,625.00
GSAATIFS333	Riot Training Suit with Accessories	4	$638.50	$2,554.00
GSAMONUTB	Riot Training Bag	10	$80.00	$800.00
BNCCELLS	C-Cell Batteries	0	$1.00	$0.00
BNAAACELLS	AA Batteries	0	$0.75	$0.00
GSASFDL123A	3V Lithium Batteries	0	$2.85	$0.00
BN9V	9V Batteries	0	$2.00	$0.00
ATIPMI	Packaging Marking and Integration Fee			$20,300.00
	Hardware Tools		**Item Total**	**$125,683.25**
	U.S. Army NLW Capability Set	5	**$125,683.25**	**$628,416.25**

2. USMC Nonlethal Weapons Capability Set

a. USMC NLW Company-Size Set

Table E-2 U.S. Marine Corps FY02 Company-Size Nonlethal Capability Set

TAMCN C41202E	*NSN 9999-01-464-2285*	*SET A (Company Size)*		
NSN/Product ID	***Name***	***Quantity***	***Unit cost***	***Item Total***
GSAATINCSPMDK5	Non-Ballistic USMC Riot Face Shields (S/R)	200	$42.50	$8,500.00
GSAATIPMBS9	Non-Ballistic Riot Body Shields Paulson	60	$88.50	$5,310.00
GSAATIUSMCSG	Non-Ballistic Riot Shin Guards	200	$60.00	$12,000.00
GSAATI23/36	USMC 23"/36" Riot Baton (Expandable) (w/Holster)	200	$73.25	$14,650.00
GSAATITRAINING BATON	USMC Riot Baton (Practice) (Price TBD)	200	$20.00	$4,000.00
GSAATIA2M	Portable Bullhorn	12	$82.00	$984.00
GSAATIDC/W	USMC Disposable Handcuffs	400	$0.94	$376.00
GSAATIDC/T	USMC Disposable Handcuff Training Device	100	$0.94	$94.00
GSAATIEAMK4	MK-4 Pouch OD	200	$6.75	$1,350.00
GSADEF5049	MK-4 Individual Live (OC) Dispenser (Pepper Spray)	200	$7.29	$1,458.00
GSADEF5099	MK-9 Live Squad (OC) Dispenser (Pepper Spray)	40	$26.20	$1,048.00
GSATIEAMK9	MK-9 Pouch OD	40	$11.00	$440.00
GSAATIDEF5846L	MK-46 Live (OC) (Horizontal) (Pepper Spray)	12	$300.00	$3,600.00
GSAATIDEF1515	MK-46 Sling	12	$30.25	$363.00
GSAATIDEF5520	MK-46 Hardware Kit	4	$68.50	$274.00
GSAATIDEF5546	MK-46 Refill Unit (OC) Live (Pepper Spray)	4	$46.25	$185.00
GSADEF5149	MK-4 Inert Individual Training (OC) Dispenser	200	$6.07	$1,214.00
GSADEF5199	MK-9 Inert Squad Training (OC) Dispenser	40	$20.10	$804.00
GSAATIDEF5946L	MK-46 Inert Training (OC) (Horizontal) (w/Dye)	12	$240.00	$2,880.00
GSADEF5546I	MK-46 Refill Unit (OC) Inert Training (w/Dye)	4	$27.50	$110.00
GSAATIFS333	Riot Training Suit w/Accessories (F.I.S.T.)	12	$638.50	$7,662.00
GSAMONUTB	Riot Training Strike Bag	24	$80.00	$1,920.00
GSAATIRSC3	Caltrop	400	$10.50	$4,200.00
TBD	Stinger Road Side Spike Kit (15.5') (Price TBD)	12		
XENNLCSUSMC	High Intensity Light (Xenonics)	6	$2,975.00	$17,850.00
GSAATIEATAVLFP6POD	Eagle Sure Fire 6P Holster	60	$8.75	$525.00
GSASF6P	Individual High Intensity Light	60	$42.74	$2,564.40
GSASFP60	Individual High Intensity Light Accessory Kit	30	$12.53	$375.90
GSASFDL123A	3V Lithium Batteries	120	$2.85	$342.00
GSAATISSSCQTR	12-Gauge Buttstock Cuff	36	$10.75	$387.00
GSAATIEAUPMS	12-Gauge Utility (25-Round) Pouch	72	$23.65	$1,702.80
GSAATIEA40MM6	40-mm Carrying Pouch	36	$13.65	$491.40
GSAATISASDD60D	Diversionary/Sting Ball Grenade Pouch	60	$47.25	$2,835.00
CSI9512	12-Gauge Launching Cup (Availability and Price TBD)	36		
1005-01-371-5639/ 1005-01-251-8578	12-Gauge Mossberg 500A2 / 590 Shotgun	36	$250.00	$9,000.00
ATIPMI	Packaging Marking and Integration Fee			$11,578.00
	Hardware Totals		**Item Total**	**$121,073.50**

b. USMC NLW Platoon-Size Set

Table E-3 USMC FY02 NLW Capability Set—Platoon Size

TAMCN C41202E	***NSN 9999-01-464-2285***	***SET B (Platoon Size)***		
NSN/Product ID	***NAME***	***Quantity***	***Unit Cost***	***Item Total***
GSAATINCSPMDK5	**Non-Ballistic USMC Riot Face Shields (S/R)	50	$42.50	$2,125.00
GSAATIPMBS9	Non-Ballistic Riot Body Shields Paulson	15	$88.50	$1,327.50
GSAATIUSMCSG	Non-Ballistic Riot Shin Guards	50	$60.00	$3,000.00
GSAATI23/36	USMC 23"/36" Riot Baton (Expandable) (w/Holster)	50	$73.25	$3,662.50
GSAATITRAINING BATON	USMC Riot Baton (Practice) (Price TBD)	50	$20.00	$1,000.00
GSAATIA2M	Portable Bullhorn	3	$82.00	$246.00
GSAATIDC/W	USMC Disposable Handcuffs	100	$0.94	$94.00
GSAATIDC/T	USMC Disposable Handcuff Training Device	25	$0.94	$23.50
GSAATIEAMK4	MK-4 Pouch OD	50	$6.75	$337.50
GSADEF5049	MK-4 Individual Live (OC) Dispenser (Pepper Spray)	50	$7.29	$364.50
GSADEF5099	MK-9 Live Squad (OC) Dispenser (Pepper Spray)	10	$26.20	$262.00
GSATIEAMK9	MK-9 Pouch OD	10	$11.00	$110.00
GSAATIDEF5846L	MK-46 Live (OC) (Horizontal) (Pepper Spray)	3	$300.00	$900.00
GSAATIDEF1515	MK-46 Sling	3	$30.25	$90.75
GSAATIDEF5520	MK-46 Hardware Kit	2	$68.50	$137.00
GSAATIDEF5546	MK-46 Refill Unit (OC) Live (Pepper Spray)	2	$46.25	$92.50
GSADEF5149	MK-4 Inert Individual Training (OC) Dispenser	50	$6.07	$303.50
GSADEF5199	MK-9 Inert Squad Training (OC) Dispenser	10	$20.10	$201.00
GSAATIDEF5946L	MK-46 Inert Training (OC) (Horizontal) (w/Dye)	3	$240.00	$720.00
GSADEF5546I	MK-46 Refill Unit (OC) Inert Training (W/Dye)	2	$27.50	$55.00
GSAATIFS333	Riot Training Suit W/Accessories (F.I.S.T.)	3	$638.50	$1,915.50
GSAMONUTB	Riot Training Strike Bag	3	$80.00	$240.00
GSAATIRSC3	Caltrop	50	$10.50	$525.00
TBD	Stinger Road Side Spike Kit (15.5') (Price TBD)	3		
XENNLCSUSMC	High Intensity Light (Xenonics)	4	$2,975.00	$11,900.0 0
GSAATIEATAVLFP6POD	Eagle Sure Fire 6P Holster	50	$8.75	$437.50
GSASF6P	Individual High Intensity Light	50	$42.74	$2,137.00
GSASFP60	Individual High Intensity Light Accessory Kit	25	$12.53	$313.25
GSASFDL123A	3V Lithium Batteries	100	$2.85	$285.00
GSAATISSSCQTR	12-Gauge Buttstock Cuff	15	$10.75	$161.25
GSAATIEAUPMS	12-Gauge Utility (25 Round) Pouch	30	$23.65	$709.50
GSAATIEA40MM6	40-mm Carrying Pouch	50	$13.65	$682.50
GSAATISASDD60D	Diversionary/Sting Ball Grenade Pouch	50	$47.25	$2,362.50
CSI9512	12-Gauge Launching Cup (Availability and Price TBD)	15		
1005-01-371-5639/ 1005-01-251-8578	12-Gauge Mossberg 500A2 / 590 Shotgun	15	$250.00	$3,750.00
ATIPMI	Packaging Marking and Integration Fee			$3,000.00
	Hardware Totals		**Item Total**	**$43,471.25**

** Riot shield. Marine Corps security forces are currently evaluating to use in conjunction with each other as listed below. They are not currently deployed but are undergoing field evaluation.
(1) The six feet Armadillo Interlocking Shield for the hard baseline.
(2) The four feet.. Armadillo Interlocking Shield for the soft baseline.

3. USN NLW Capability Set

a. The USN is in the process of developing NLW capabilities and capability sets.* The sets will support the specific areas of operation, mission, and/or user.

b. The first version will address the force protection mission and will be based on other service-approved items and some longer-range commercial-off-the-shelf items currently being upgraded for military use. The mobile security forces, master at arms, and shipboard force protection units are the targeted users.

c. A notional capability set for the Mobile Security Forces is listed below.

Table E-4. Notional USN Capability Set

Name	*Nomenclature (Mfr. Representative)*	*Quantity*	*Unit Cost*	*Case Size*	*Cost Extended*
Fin Stabilize Round		1 Case	$5	100	$400
Stingshot		1 Case	$5	100	$400
Shotbag		1 Case	$5	100	$450
Flamethrower Round		1 Case	$8	100	$750
Tactical Blast Bursting Round		1 Case	$7	100	$650
Ultraflash Bursting Round		1 Case	$8	100	$750
Launchable Stun Munition	POP, Inc. T-444	1 Case	$60	24	$1,440
40-mm Sparkshot		1 Case	$35	24	$840
40-mm Stingshot		1 Case	$25	24	$600
40-mm Shotbag		1 Case	$25	24	$600
40-mm OC		1 Case	$28	24	$660
Stinger	POP, Inc. I-452	1 Case	$60	24	$1,440
Stun	POP, Inc. T-470	1 Case	$55	24	$1,320
Flash/Bang Multi	POP, Inc. D-451	1 Case	$65	24	$1,560
OC	POP, Inc. I-452	1 Case	$65	24	$1,560
12-Gauge Launching Blanks		200	$3		$600
Grenade Training Rounds		12	$65		$780
12-Gauge Training Rounds		24	$3		$72
40-mm Training Rounds		12	$20		$240
12-Gauge Shotguns	Standard Issue/Modified	12	$1,000		$12,000
12-Gauge Launching Adapter		6	$325		$1,950
Face Shields		15	$100		$1,500
Helmets		15	$200		$3,000

* In FY 01, the USN fielded an USMC capability set to Naval Air Station, Roosevelt Roads, Puerto Rico.

Table E-4. Notional USN Capability Set

Name	*Nomenclature (Mfr. Representative)*	*Quantity*	*Unit Cost*	*Case Size*	*Cost Extended*
Tactical Vest		15	$300		$4,500
Bullhorn		2	$100		$200
Medical Kit		2	$50		$100
Searchlights		2	$100		$200
Field Pack		15	$100		$1,500
Holsters/Belts		15	$50		$750
Ordnance Packaging, Handling, Storage and Transportation (PHS&T)					$1,500
Misc. (Certification, Training and Documentation)					$3,000
			Total Cost Basic Kit		**$45,312**

Table E-5. Notional USN Capability Set

Name	*Nomenclature (Mfr. Representative)*	*Quantity*	*Unit Cost*	*Case Size*	*Cost Extended*
Stingmore Mine (Std Body) with Rubber Ball Projectiles	POP, Inc. I-454	24	$350		$8,400
Stingmore Mine Stand		6	$100		$600
Electric Firing Boxes		8	$35		$280
Area Lighting					$1,000
Electric Generator					$500
Signs and Placards					$500
Personal NVGs and Opticals		6	$2,000		$12,000
Additional PHS&T					$2,000
Misc. (Certification, Training and Documentation)					$2,000
Total Cost Augmented Basic Kit					**$27,280**
Total Cost (Basic plus Augment)					$72,592
Portable Vehicle Arresting Barrier	U.S. Army	1	$40,000		$40,000
Running Gear Entanglement System (RGES)	NSWCDD	1000 ft system	$25,000		$25,000
Antimaterial Round (USCG)		24	$12		$288
Spike Strips		2	TBD		
Improved Sensor (s)					
RF			TBD		

Table E-5. Notional USN Capability Set

Name	Nomenclature (Mfr. Representative)	Quantity	Unit Cost	Case Size	Cost Extended
IR			TBD		
Acoustic			TBD		
Chemical Detection	SABRE 2000	1	$20,000		$20,000
Ordnance Packaging, Handling, Storage and Transportation (PHS&T)					$6,000
Misc. (Certification, Training and Documentation)					$2,000
Total Cost Vehicle Supplement					**$93,288**
Total Cost Basic, Augmented, and Vehicle Supplement					**$165,880**

4. USAF NLW Capability Set

Table E-6. USAF NLW Capability Set

NSN	Item Description	EA	Item $	Item Total
GSADEF5149	Inert Individual (OC) Training Dispenser	1	$6.07	$6.07
GSADEF5846H	Mid-Size Riot Control Dispenser (Chemical)	3	$318.75	$956.25
GSADEF5546	Riot Control Dispenser (OC) Agent Quarts	1	$46.25	$46.25
	12-Gauge Shotgun (Area CDC/Rubber Pellet)	36	$4.46	$160.56
	12-Gauge Shotgun (Point/Fin Stabilized)	36	$2.85	$102.60
CSI (USMC) GG04	Stun Hand (Sting Ball) Grenade	50	$35.00	$1,750.00
M84	Stun Hand Grenade/Flash-Bang Device	50	$78.00	$3,900.00
GSAPMDK5X/AF	Non-Ballistic Face Shields	13	$29.50	$383.50
GSAPMBS9	Non-Ballistic Full Length Riot Shield	13	$88.50	$1,150.50
GSAUSMCSG	Non-Ballistic Shin guards	13	$6.000	$780.00
GSAATI23/36	36" Riot Baton w/Holster	13	$73.25	$952.25
GSAA2	Portable Megaphone	1	$82.00	$82.00
EAAFCUSTOM	Tactical Vest Rough Estimate	3	$150.00	$450.00
HARG700	Riot Control Gloves	13	$52.20	$678.60
Champro	Protective Cups	13	$7.25	$94.25
	40-mm (Point) Sponge Round	54	$23.00	$1,242.00
	40-mm (Area) Crowd Dispersal Cartridge Grenade	54	$22.00	$1,188.00
GSAF333	Riot Baton Training Suit	1	$638.50	$638.50

Table E-6. USAF NLW Capability Set

NSN	*Item Description*	*EA*	*Item $*	*Item Total*
MONDC/W	Disposable Forearm/Ankle Cuffs (FLEX)	40	$0.94	$37.60
GSADEF5049/ EA	Individual (OC) Dispenser with Carrier	13	$7.29	$94.77
	12-Gauge Utility Pouch	2	$24.00	$48.00
	12-Gauge Buttstock Cuff	3	$11.00	$33.00
	40-mm Carrying Pouch	2	$14.00	$28.00
	Training Strike Bag	2	$80.00	$160.00
	Cost of Clip-Lok Boxes			$480.00
	Approximate freight shipped via commercial freight into CONUS			$1,500.00
			Total Estimate	**$16,942.70**
Proposal to add TASERS to USAF Capability Set after testing.				

5. Notional USCG NLW Capability Set

Table E-7. Notional USCG NLW Capability Set

Reference: (a) COMDTINST M16247.1B Maritime Law Enforcement Manual (MLEM) (b) Project Ericson Website	
Expandable Baton	
Restraint Devices	Hand Cuffs, Flex Cuffs, Leg Irons, Migrant Restraint Harness
Individual OC Spray	
Non-Ballistic Riot Face Shield	
Non-Ballistic Riot Body Shield	
Non-Ballistic Riot Shin Guard	
Running Gear Entanglement Systems	

The following information describes each of the components of the NLW capability set currently deployed by the USCG. (The USCG has not designated a standard set of nonlethal gear. The using unit determines the gear needs.)

6. Restraint Devices

a. Handcuffs are the preferred restraint devices when dealing with a small number of subjects. Handcuffs should be steel, chain-linked construction capable of being double locked, with nickel, blued, or stainless steel finish.

b. Using flexible handcuffs for the temporary restraint of subjects is also authorized. Since they do not require a key and are relatively easy to apply, flexible handcuffs are particularly useful and appropriate when dealing with large numbers of subjects (e.g. migrants).

c. Leg irons are authorized restraint devices and may be a preferred option for restraining subjects for extended periods of time.

d. Migrant restraint harnesses are authorized when dealing with large numbers of subjects.

7. Expandable Baton

The expandable baton is the only intermediate weapon authorized for USCG personnel to carry. It should have a steel shaft that is silver in color and corrosion resistant (e.g. chrome, nickel, stainless steel) and have no knurling or sharp edges. It should have a foam or similar, non-slip grip. It should be rigid when deployed and must be no less than 15 inches and no greater than 22 inches when expanded. Spring loaded or other trigger style batons are not authorized.

8. OC Pepper Spray

Only chemical irritants containing the active ingredient OC are authorized for USCG personnel to carry. The aerosol must contain no greater than 10 percent and no less than 5 percent OC concentration. It shall have a water-based carrier, be nonflammable, nontoxic and produce a straight stream delivery (not fog).

9. Helmet Riot Face Shield

Comparable to DOD type. Not currently deployed. Undergoing field evaluation.

10. Riot Shield

Two shields are currently being evaluated. A 24 x 48 inches clear lexan weighing 14 pounds and a smaller version measuring 20 x 36 inches. They are not currently deployed. Undergoing field evaluation.

REFERENCES

Joint

JP 1-02, *DOD Dictionary of Military and Associated Terms*, 12 Apr. 2001

JP 3-07, *Joint Doctrine for Military Operations Other Than War*, 16 Jun. 1995

JP 3-07.3, *JTTP for Peace Operations*, 12 Feb.1999

JP 3-57, *Doctrine for Joint Civil-Military Operations*, 8 Feb. 2001

JP 3-58, *Joint Doctrine for Military Deception*, 31 May 1996

JP 3-61, *Doctrine for Public Affairs in Joint Operations*, 14 May 1997

JP 4-0, *Joint Doctrine for Logistic Support of Joint Operations*, 6 Apr. 2000

Multi-Service

FM 3-100.12, MCRP 5-12.1C, NTTP 5-03.5, AFTTP(I) 3-2.34, *Risk Management* for *Multiservice Tactics, Techniques, and Procedures*, 15 Feb. 2001

FM 3-31.1, MCWP 3-36, *Army and Marine Corps Integration in Joint Operations*, 21 Nov. 2001

Army

AR 190-14, *Military Police Carrying of Firearms and Use of Force for Law Enforcement and Security Duties*, 12 Mar. 1993

AR 350-38, *Training Devices Policies and Management*, 15 Oct. 1993

AR 27-1, *Judge Advocate Legal Services*, 30 Sep. 1996

FM 3-0, *Operations*, 14 Jun. 2001

FM 3-05.30 (33-1), *Psychological Operations*, 19 Jun. 2000

FM 19-15, *Civil Disturbances*, 25 Nov. 1985

FM 27-10, *The Law of Land Warfare*, 18 Jul. 1956

FM 100-19, *Domestic Support Operations*, 1 Jul. 1993

TC 90-1, *Training for Urban Operations*, 1 Apr. 2002

TRADOC Pamphlet 525-5, *Force XXI Operations, A Concept for the Evolution of Full Dimensional Operations for the Strategic Army of the Twenty-First Century*, 1 Aug. 1994

TRADOC Pam 350-9, *TRADOC Training Devices for Army wide Use*, 30 Sep. 1993

TRADOC Regulation 350-70, *Systems Approach to Training Management, Processes, and Products*, 9 Mar. 1999.

TRADOC Regulation 385-2, chg.1, TRADOC Safety Program, 10 Oct. 2000.

Marine Corps

MCWP 3-33.4, *Domestic Support Operations*, Jul. 1993

I MEF FMF, After Action Report—UNITED SHIELD, 7 Jun. 1995

MCO 1510.112, *Individual Training Standards for Nonlethal Weapons*, 27 May 1998

MCO 5500.6F, *Arming of Security and Law Enforcement Personnel and the Use of Force*, 20 Jul. 1995

MCRP 5-12.1A, *The Law of Land Warfare.*

MCO 1630.4A, *Law Enforcement and Physical Security Activities*, 19 Jun. 1991.

MCO 1510.114, *Individual Training Standards (ITS) for Antiterrorism Force Protection (AT/FT) System*, 30 Jan. 1998.

MCWP 3-11.1, *Marine Rifle Company*, Feb. 1978

MCRP 3-11.1A, *Commander's Tactical Handbook*, 30 Jul. 1988. .

MCWP 3-11.2 *Marine Rifle Squad*

Marine Corps Gazette, *"Less-Lethal" Force in Operation UNITED SHIELD*, Sep. 1995.

Navy

Naval Doctrine Publication 1, Naval Warfare (U), 28 Mar. 1994

NWP 1-14M, *The Commander's Handbook on the Law of Naval Operations*, Oct. 1995

SECNAVINST 5500.29B, *Use of Deadly Force and the Carrying of Firearms by Personnel of the Department of the Navy in Conjunction with Law Enforcement Security Duties and Personal Protection*, 29 Sep. 1992

Air Force

AFI 31-201, *Security Police Standards and Procedures*, 4 Dec. 2001

AFI 31-207, *Arming and Use of Force by Air Force Personnel*, 1 Sep. 1999

AFMAN 31-201V6, *Civil Disturbance*, 17 May 2002

AFMAN 31-222, *Security Forces Use of Force Manual*, 1 Jun. 2001

Coast Guard

COMDTINST M16247.1B, Maritime Law Enforcement Manual, 9 Aug. 2001

Other

DODD 3000.3-R, *Policy for Nonlethal Weapons*, 9 Jul. 1996

DODD 3025.1, *Military Support to Civil Authorities (MSCA)*, 15 Jan. 1993

DODD 5100.77, *DOD Law of War Program*, 9 Dec. 1998

DODD 3025, *I-M Manual for Civil Emergencies*, 2 Jun. 1994

DODD 3025.12, *Military Assistance for Civil Disturbances*, 4 Feb. 1994

DODD 3025.13, *DOD Resources in Support of US Secret Service*, 13 Sep. 1985

DODI 5000.2, *Operation of the Defense Acquisition System*, 5 Apr. 2002

DODR 5000.1, *Defense Acquisition*, 23 Oct. 2000, Ch 1, 4 Jan. 2001

DOD Joint Concept for Nonlethal Weapons, 5 Jan. 1998

DODD 3025.12, *Military Assistance for Civil Authorities*, 18 Feb. 1997

DODD 3025.16, *Military Assistance for Civil Authorities Liaison Office (EPLO) Program,*18 Dec. 2000

DODI 5000.2, *Operation of the Defense Acquisition System*, 5 Apr. 2002

Mission Need Statement (MNS), *Range of Nonlethal Force Capabilities (NLFC)*, 10 Oct. 1996

A Guide for Effective Peacetime Employment of Reserve Components Units and Individuals, OASD RA (M&P), Dec. 2000

Defense and Intelligence Abbreviations and Acronyms, Nov. 1997

Emergency Management Assistance Compact—Guide Book and Operating Procedures, National Emergency Management Association, Aug. 2000

Inside the Navy, Marine Corps Nonlethal Weapons Experiment Underway at Camp Pendleton, CA, Vol. 9, No. 29, Jul. 22, 1996

Military Support To Civil Authorities—Interagency Disaster Resource Guide, National Interagency Civil-Military Institute, Jan. 2001

National Guard Bureau Military Support to Civil Authorities Lessons Learned, 6 Sep. 2000

The Robert T. Stafford Disaster Relief and Emergency Assistance Act, as amended 42 USC 1521, Jul.2001

Joint Universal Lessons Learned System, Operation UNITED SHIELD

Nonlethal Weapons Joint Mission Area Analysis/Joint Mission Needs Analysis, Dec. 2000

Interservice Nonlethal Individual Weapons Instructor Course (INIWIC) Program of Instruction, Oct.2000

Presidential Decision Directive PDD-39 U.S. Policy on Counterterrorism 21 Jun.1995

PDD-62 Combating Terrorism—Fact Sheet, 22 May 1998

The Federal Response Plan Notice of Change 7, Feb.1997; FEMA, Number 229 Change 11

Federal Response Plan, Apr 1999

GLOSSARY

PART I – ABBREVIATIONS AND ACRONYMS

A	
ACOG	advanced combat optical gunsight
ADS	aerial delivery system
AFDC	Air Force Doctrine Center
AFI	Air Force instruction
AFMAN	Air Force manual
ALSA	Air Land Sea Application
AMC	Army Materiel Command
AO	area of operation
AR	Army regulation
ATSC	Army Training Support Center
B	
bldg	building
BLT	battalion landing team
bn	battalion
C	
C4ISR	command, control, communications, computers, intelligence, surveillance, and reconnaissance
CA	civil affairs
cal	caliber
CANS	compressed air net system
CDC	crowd disposal cartridge
CMC	Commandant of the Marine Corps
COB	close of business
COC	combat operations center
COMDTPUB	Commandant, Coast Guard Publication
COMDTINST	Commandant, Coast Guard Instruction
CONUS	continental United States
COTS	commercial off-the-shelf
CP	command post
CQB	close quarters battle
CS	O-Chlorobenzyl-Malononitrile (riot control agent (tear gas))
CSSOC	combat service support operations center
cu ft	cubic feet

D	
DA	Department of the Army
dB	decibel
DECON	decontamination
demo	demonstration
DM	designated marksman
DOD	Department of Defense
DODD	Department of Defense Directive
DODI	Department of Defense Instruction
DODIC	Department of Defense Identification Code
DODAC	Department of Defense Ammunition Code
DOT	Department of Transportation
DP	downed person
DPB	deployable pursuit boat
DTD	dated
E	
EADBH	Eagle Duty Baton Holder
EEZ	exclusive economic zone
EOD	explosive ordnance disposal
F	
FAST	Fleet Antiterrorism Security Team
FDP	food distribution point
FLEX	forearm/ankle cuff
FM	field manual
four M's	mind, mass, movement, muscle
FSE	fire support element
ft	feet
G	
G1	Army or Marine Corps component manpower or personnel staff officer (Army division or higher staff, Marine Corps brigade or higher staff)
G2	Army or Marine Corps component intelligence staff officer (Army division or higher staff, Marine Corps brigade or higher staff)
G3	Army or Marine Corps component operations staff officer (Army division or higher staff, Marine Corps brigade or higher staff)

G4	Army or Marine Corps component logistics staff officer (Army division or higher staff, Marine Corps brigade or higher staff)
ga	gauge
GOTS	government off-the-shelf
H	
HMMWV	high mobility multipurpose wheeled vehicle
I	
IADS	Integrated Air Defense System
IAW	in accordance with
ID	identify/identification
IFF	identification, friend or foe
in.	inch
INIWIC	Interservice Nonlethal Individual Weapons Instructor Course
IPB	intelligence preparation of the battlespace
ITRO	Interservice Training Organization
J	
JMAA	Joint Mission Area Analysis
JTTP	joint tactics, techniques, and procedures
K	
KFOR	Kosovo forces
L	
LBE	load-bearing equipment
LOAC	Law of Armed Conflict
LP	listening post
LVOSS	light visual optical screening system
M	
M16	M16A2 rifle
M203	M203 Grenade Launcher
MARCORSYCOM	Marine Corps System Command
MCCDC	Marine Corps Combat Development Command
MCCM-GE	modified crowd control munition—ground emplaced
MCO	Marine Corps Order
MCPDS	Marine Corps Doctrinal Publication System
MCRP	Marine Corps reference publication

MCSF	Marine Corps security force
MCWP	Marine Corps warfighting publication
MDST	Monadnock Defensive Tactics System
MEDEVAC	medical evacuation
MEF	Marine expeditionary force
METT-T	mission, enemy, terrain and weather, troops and support available – time available
METT-TC	mission, enemy, terrain and weather, troops and support available – time available and civil considerations
MEU	Marine expeditionary unit
MIO	Maritime interception operation
MLEM	Maritime Law Enforcement Manual
MILSTRIP	military
mm	millimeter
MNS	mission need statement
MOAB	management of aggressive behavior
MOOTW	military operations other than war
MOS	military occupational specialty
MOUT	military operations in urban terrain
MP	military police
mph	miles per hour
MRE	meal, ready to eat
MTTP	multi-Service tactics, techniques, and procedures

N

NATO	North Atlantic Treaty Organization
NAVSOP	Navy standing operating procedures
NDC	Naval Doctrine Command
NLFC	nonlethal force capabilities
NLW	nonlethal weapon
NSN	national stock number
NSWC	Naval Surface Warfare Command
NWDC	Naval Warfare Development Command
NWP	Naval Warfare Publication

O

OC	Oleoresin Capsicum
OCAT	Oleoresin Capsicum aerosol training
OCONUS	outside the continental United States
OD	olive drab
OP	observation post

OPFOR	opposing force
P	
PA	public affairs
PAO	public affairs officer
PASGT	personnel armor system, ground troops
POI	program of instruction
PSYOP	psychological operations
PT	physical training
PVAB	portable vehicle arresting barrier
PVC	polyvinyl chloride
Q	
QRF	quick reaction force
R	
R&D	research and development
RCA	riot control agent
RGES	running gear entanglement system
ROE	rules of engagement
RRF	rapid reaction force
S	
S1	battalion or brigade manpower or personnel staff officer (Army; Marine Corps battalion or regiment)
S2	battalion or brigade intelligence staff officer (Army; Marine Corps battalion or regiment)
S3	battalion or brigade operations staff officer (Army; Marine Corps battalion or regiment)
S4	battalion or brigade logistics staff officer (Army; Marine Corps battalion or regiment)
SASO	stability and support operations
SB	shield bearer
sec	second
SECNAVINST	Secretary of the Navy instruction
SG	shotgun
SJA	staff judge advocate
SNCO	staff noncommissioned officer
SOP	standard operating procedure
SOTG	Special Operations Training Group
SROE	standing rules of engagement
SSC	small-scale contingency

T

TC	training circular
TF	task force
T/O	table of organization
TOC	tactical operations center
TRADOC	United States Army Training and Doctrine Command
TTP	tactics, techniques, and procedures

U

UN	United Nations
US	United States
USA	United States Army
USAF	United States Air Force
USAMPS	United States Army Military Police School
USCG	United States Coast Guard
USMC	United States Marine Corps
USN	United States Navy
USS	United States Ship

V

VCR	videocassette recorder
VLNL	vehicle launched nonlethal grenade
VMNLGL	vehicle mounted nonlethal grenade launcher

W

w/	with
WMD	weapons of mass destruction

PART II – TERMS AND DEFINITIONS

Air Force specialty code (AFSC). Primary job of a U.S. Air Force officer or enlisted person. Analogous to the Army and Marine Corps military occupational skill (MOS).

area of operation (AO). An operational area defined by the joint force commander for land and naval forces. Areas of operation do not typically encompass the entire operational area of the joint force commander, but should be large enough for component commanders to accomplish their missions and protect their forces. (JP 1-02)

battalion landing team (BLT). In an amphibious operation, an infantry battalion normally reinforced by necessary combat and service elements; the basic unit for planning an assault landing. (JP 1-02)

civil affairs. Designate active and Reserve Component forces and units organized, trained, and equipped specifically to conduct civil affairs activities and to support civil-military operations. (JP 1-02)

civil-military operations. The activities of a commander that establish, maintain, influence, or exploit relations between military forces, governmental and nongovernmental civilian organizations and authorities, and the civilian populace in a friendly, neutral, or hostile operational area in order to facilitate military operations, to consolidate and achieve operational U.S. objectives. Civil-military operations may include performance by military forces of activities and functions normally the responsibility of the local, regional, or national government. These activities may occur prior to, during, or subsequent to other military actions. They may also occur, if directed, in the absence of other military operations. Civil-military operations may be performed by designated civil affairs, by other military forces, or by a combination of civil affairs and other forces. (JP 1-02)

combat service support. The essential capabilities, functions, activities, and tasks necessary to sustain all elements of operating forces in theater at all levels of war. Within the national and theater logistics systems, it includes, but is not limited to, that support rendered by service forces in ensuring the aspects of supply, maintenance, transportation, health services, and other services required by aviation and ground combat troops to permit those units to accomplish their missions in combat. Combat service support encompasses those activities at all levels of war that produce sustainment to all operating forces on the battlefield. (JP 1-02)

command and control. The exercise of authority and direction by a properly designated commander over assigned and attached forces in the accomplishment of the mission. Command and control functions are performed through an arrangement of personnel, equipment, communications, facilities, and procedures employed by a commander in planning, directing, coordinating, and controlling forces and operations in the accomplishment of the mission. (JP 1-02)

commercial off the shelf (COTS). Describes items procured for military use through civilian contractors.

core capability. The set of functions that define a significant, stand-alone, operationally effective and suitable military capability such that, should no further development occur, the user will have received a significant capability.

designated marksman (DM). A rifleman stationed in an overwatch position above friendly forces, armed with a standard infantry weapon appropriate to his service that is mounted with a high-powered scope and loaded with lethal ammunition. The designated marksman uses his vantage point to identify agitators to forces below, eliminate lethal threats, and monitor the battlefield.

dispersion. A scattered pastern of hits around the mean point of impact of bombs and projectiles dropped or fired under identical conditions. In chemical and biological operations, the dissemination of agents in liquid or aerosol form. (JP 1-02)

fleet Marine force (FMF). A balanced force of combined arms comprising land, air, and service elements of the U.S. Marine Corps. A fleet Marine force is an integral part of a U.S. Fleet and has the status of a type command. (JP 1-02)

force protection. Security program designed to protect service members, civilian employees, family members, facilities, and equipment, in all locations and situations, accomplished through planned and integrated application of combating terrorism, physical security, operations security, and personal protective services, and supported by intelligence, counterintelligence, and other security programs. (JP 1-02)

government off-the-shelf (GOTS). Analogous to COTS, but describing items already procured by the U.S. government and ready for issue to military units.

host nation. A nation that receives the forces and/or supplies of allied nations, coalition partners, and/or NATO organizations to be located on, to operate in, or to transit through its territory. (JP 1-02)

intelligence preparation of the battlespace (IPB). An analytical methodology employed to reduce uncertainties concerning the enemy, environment, and terrain for all types of operations. Intelligence preparation of the battlespace builds an extensive database for each potential area in which a unit may be required to operate. The database is then analyzed in detail to determine the impact of the enemy, environment, and terrain on operations, and present it in graphic form. Intelligence preparation of the battlespace is a continuing process. Also called IPB. (JP 1-02)

interrogator-translator team (ITT). Marine Corps unit, usually working as a small detachment to a larger infantry unit, highly trained in interrogation techniques and fluent in adversary native language.

joint force commander. A general term applied to a combatant commander, subunified commander, or joint task force commander authorized to exercise combatant command (command authority) or operational control over a joint force. (JP 1-02)

logistics. The science of planning and carrying out the movement and maintenance of forces. In its most comprehensive sense, those aspects of military operations which deal with:

a. Design and development, acquisition, storage, movement, distribution, maintenance, evacuation, and disposition of materiel.

b. Movement, evacuation, and hospitalization of personnel.

c. Acquisition of construction, maintenance, operation, and disposition of facilities.

d. Acquisition of furnishing of services. (JP 1-02)

low-intensity conflict. Political-military confrontation between contending states or groups below conventional war and above the routine, peaceful competition among states. It frequently involves protracted struggles of competing principles and ideologies. Low intensity conflict ranges from subversion to the use of armed force. It is waged by a combination of means employing political, economic, informational, and military instruments. Low-intensity conflicts are often localized, generally in the Third World, but contain regional and global security implications.

NOTE: Also known as small wars or "brushfire wars," low-intensity conflicts are rebellions, insurrections, and the like characterized by adversaries dedicated to a cause but lacking command and control structure and modern equipment.

Marine expeditionary force (MEF). The largest of the Marine air-ground task force (MAGTF) and the Marine Corps principal warfighting organization, particularly for larger crisis or contingencies. It is task-organized around a permanent command element and normally contains one or more Marine divisions, Marine aircraft wings, and Marine force service support groups. The Marine expeditionary force is capable of missions across the range of military operations, including amphibious assault and sustained operations ashore in any environment. It can operate from a sea base, a land base, or both. Also called MEF. See also aviation combat element; combat service support element; command element; Marine air-ground task force; Marine expeditionary force (forward); Marine expeditionary unit; special purpose Marine air-ground task force; task force. (JP 1-02)

maritime interception operation (MIO). Means by which an exclusion zone at sea is enforced. Involves the visit of merchant ships, by armed boarding parties if necessary, bound to, through, or out of a defined area, to examine each ship's papers and cargo and to search for evidence of contraband. The provisions must be established by a sanctioning body and must be applied to ships of all nationalities.

military operations other than war (MOOTW). Operations that encompass the use of military capabilities across the range of military operations short of war. These military actions can be applied to complement any combination of the other instruments of national power and occur before, during, and after war. (JP 1-02)

military occupational specialty (MOS). Primary job of military officer or enlisted personnel.

military operations in urban terrain (MOUT). All military actions planned and conducted on a topographical complex and its adjacent natural terrain where man-made construction is the dominant feature. It includes combat-in-cities, which is that portion of MOUT involving house-to-house and street-by-street fighting in towns and cities.

military police (MP). Marine Corps and Army police forces designated by specific MOS.

maritime pre-positioning force operation. A rapid deployment and assembly of a Marine expeditionary force in a secure area using a combination of strategic airlift and forward-deployed maritime pre-positioning ships. See also Marine expeditionary force; maritime pre-positioning ships (JP 1-02 and JP 4-01.6)

nonlethal weapons (NLW). Weapons that are explicitly designed and primarily employed so as to incapacitate personnel or materiel, while minimizing fatalities, permanent injury to personnel, and undesired damage to property and the environment.

a. Unlike conventional weapons that destroy their targets through blast, penetration, and fragmentation, nonlethal weapons employ means other than gross physical destruction to prevent the target from functioning.

b. Nonlethal weapons are intended to have one, or both, of the following characteristics:

(1) They have relatively reversible effects on personnel or materiel.

(2) They affect objects differently within their area of influence. (JP 1-02)

nonlethal weapon capability set. A set of systems containing weapon systems, munitions, and protective equipment. The components are explicitly designed and primarily employed to provide additional protection to friendly personnel and to incapacitate hostile personnel or material, while minimizing fatalities or permanent injury to intended targets and minimal damage to property and the environment. They are not required to have zero probability of producing fatalities or permanent injuries but are designed and employed in a manner that significantly reduces those probabilities over the traditional employment of currently fielded military weapon systems, munitions, and equipment.

nonlethal technology (N-LT). Synonymous with NLW, but also encompassing more advanced technology that is not yet available for use by military forces.

Oleoresin Capsicum (OC). Also known as "pepper spray", OC is made completely from organic materials and is FDA-approved for sale over-the-counter in the United States. Causes severe and immediate burning sensation to mucous membranes when sprayed into face, nose, and eyes.

psychological operations (PSYOP). Planned operations to convey selected information and indicators to foreign audiences to influence their emotions, motives, objective reasoning, and ultimately the behavior of foreign governments, organizations, groups, and individuals. The purpose of psychological operations is to induce or reinforce foreign attitudes and behavior favorable to the originator's objectives. (JP 1-02)

rules of engagement (ROE). Directives issued by competent military authority that delineate the circumstances and limitations under which U.S. forces will initiate and/or continue combat engagement with other forces encountered. See also law of war. (JP 1-02)

table of organization (T/O). Total manpower authorized for specific unit, organized by billet and rank. TO& E is a U.S. Army term which encompasses T/O and T/E.

training aids and devices. Any item developed or procured with the primary intent to assist in training and learning.

weapons of mass destruction (WMD). Weapons capable of a high order of destruction and/or of being used in such a manner as to destroy large numbers of people. Weapons of mass destruction can be high explosives or nuclear, biological, chemical, and radiological weapons, but exclude the means of transporting or propelling the weapon where such means is a separable and deliverable part of the weapon. (JP 1-02)

Index

Foreword

We remain a nation at war in an era of persistent conflict, but we do not stand alone. Our nation has many multinational partners, equally committed to freedom, rule of law and stability. It is clear that we are stronger when we act with partners in today's operating environment. Therefore, security force assistance is no longer an "additional duty." It is now a core competency of our Army.

As distinctions separating war and peace blur and challenges to security increase, we must seek to help our multinational partners successfully confront their security challenges. Security force assistance builds our multinational partners' capability to defeat regular, irregular, and hybrid threats prevalent in an era of persistent conflict.

The two pillars of security force assistance are the modular brigade and Soldiers acting as advisors. Their tactical efforts build partner capability and improve the security situation. Over time, U.S. forces and partners foreign security forces collectively set the conditions to defeat common threats and ultimately achieve strategic success.

This manual addresses common characteristics and considerations for conducting security force assistance and clarifies what units and individual advisors must understand to work "by, with, and through" their counterparts. Since every situation and foreign security force is unique, units and individuals conducting security force assistance must carefully analyze the operational environment, especially the relationships of foreign security forces and their population. Conducting successful security force assistance requires adaptive units led by well-informed, culturally astute leaders. FM 3-07.1 provides commanders, staffs, and advisors with the common doctrine for achieving this goal.

Conducting foreign security assistance requires great skill in building relationships and "leading from behind." We must all increase our understanding of this important mission.

MARTIN E. DEMPSEY
General, U.S. Army
Commanding General
U.S. Army Training and Doctrine
Command

Field Manual
No. 3-07.1

Headquarters
Department of the Army
Washington, DC, 1 May 2009

Security Force Assistance

Contents

Distribution Restriction: Approved for public release; distribution is unlimited.

Figures

Preface

This field manual (FM) is the Army's doctrinal publication for security force assistance (SFA). It provides doctrinal guidance and direction for how U.S. forces contribute to SFA. It focuses on the brigade combat team (BCT) conducting SFA and advising foreign security forces. It is based on lessons learned from previous advising efforts and recent combat operations with a view to the future. It supports the Army Education System instruction on the theory and conduct of SFA.

The two primary audiences for this manual are leaders in BCTs conducting SFA and Soldiers assigned as advisors. The BCT is the largest fixed tactical organization and the key formation of the Army's modular organization. Army modularity allows commanders to add selective units to assist the BCT as it conducts SFA. The BCT and higher echelons provide the framework for advisors to function and accomplish the mission—developing foreign security forces and, when appropriate, working by, with, and through foreign security forces to achieve the desired end state of the host nation's internal defense and development. Thus, Soldiers assigned as advisors are the key element of SFA.

This FM provides the conceptual framework for conventional forces to conduct SFA within the construct of full spectrum operations, across the spectrum of conflict. It addresses SFA at operational and tactical levels.

Army doctrine is consistent and compatible with joint doctrine. FM 3-07.1 expands on fundamental principles for SFA expressed in FM 3-07. FM 3-07.1 uses text and concepts developed with North Atlantic Treaty Organization partners and ABCA (America, Britain, Canada, Australia, and New Zealand program armies).

Terms that have joint or Army definitions are italicized and the number of the proponent manual follows the definition. FM 3-07.1 will be the proponent manual for foreign security forces as a term.

FM 3-07.1 applies to the Active Army, Army National Guard/U.S. Army National Guard, and U.S. Army Reserve unless otherwise stated.

Headquarters, US Army Training and Doctrine Command is the proponent for this publication. The preparing agency is the Combined Arms Doctrine Directorate, US Army Combined Arms Center. Send written comments and recommendations on Department of the Army (DA) Form 2028 (Recommended Changes to Publications and Blank Forms) directly to: Commander, US Army Combined Arms Center and Fort Leavenworth, ATTN: ATZL-CD (FM 3-07.1), 201 Reynolds Avenue (Building 285), Fort Leavenworth, KS 66027-1352. Send comments and recommendations by e-mail to leav-cadd-web-cadd@conus.army.mil. Follow the DA Form 2028 format or submit an electronic DA Form 2028.

Introduction

Our strategy emphasizes building the capacities of a broad spectrum of partners as the basis for long-term security. We must also seek to strengthen the resiliency of the international system to deal with conflict when it occurs.

National Defense Strategy, June 2008

In an era of persistent conflict, the United States supports the internal defense and development of international partners, regardless of whether those partners are highly developed and stable or less developed and emerging. While many of these partners are nations, they can also include alliances, coalitions, and regional organizations. U.S. support to these partners ranges from providing humanitarian assistance to major combat operations. U.S. support includes conducting conflict transformation, bolstering partner legitimacy, and building partner capacity. A vital part of these three aspects of U.S. support is assisting partner security forces.

Security force assistance (SFA) is not new for Army forces. In fact, General George Washington's Inspector General of the Army acted as an advisor for Army forces. Baron Friedrick Wilhelm von Steuben instilled discipline and professionalism into an army that previously lacked formalized training. His 1779 *Regulations for the Order and Discipline of the Troops of the United States*, adapted from the Prussian army, formed the doctrinal backbone of the Continental Army throughout the Revolutionary War. Additionally, the lineage of the Army's operations field manual, FM 3-0, can be traced to this document. As a benefactor of advisors such as von Steuben, the Army has since undertaken what is called SFA several times throughout its history.

Army doctrine defines *security force assistance* as the unified action to generate, employ, and sustain local, host-nation, or regional security forces in support of a legitimate authority (FM 3-07). SFA is part of the FM 3-0 construct of full spectrum operations. Similarly, it is conducted across the spectrum of conflict or in any of the operational themes. SFA is normally part of a larger security sector reform effort, while in other instances, SFA is not tied to reform but to building partner capacity.

Three general situations exist in which SFA may occur: an internally focused bilateral relationship, an externally focused bilateral relationship, and a multilateral relationship. Consequently, SFA supports the appropriate partner's plans. When SFA supports a host nation, it also supports that host nation's strategy. If SFA supports a host nation's externally focused efforts, it must support the host nation's national security strategy. SFA may support regional security forces, such as those of the African Union or the Organization of American States. In these cases, SFA supports that organization's plans.

Organization

This FM primarily focuses on the brigade combat team and advisors. It also provides a wider doctrinal framework for SFA. It is organized in the following manner:

- **Chapter 1 focuses above the brigade combat team**. It provides the overall doctrinal context for SFA.
- **Chapters 2 through 6 focus on the brigade combat team level**. These chapters provide a framework for SFA, its considerations for the brigade combat team operations process, considerations for augmenting the modular brigade for security force assistance, its considerations for the unit employment, and its sustainment considerations.
- **Chapters 7 through 10 focus on the individual advisor**. These chapters discuss the advisor, advisor cultural and communication considerations, the advisor working with counterparts, and cross-cultural influencing and communication.
- The appendices discuss Army special operations forces imperatives, legal considerations, and media considerations.

Chapter 1
The Strategic Context

This chapter places security force assistance in strategic context. First, it explains security force assistance and its relationship to full spectrum operations, spectrum of conflict, and operational themes (see FM 3-0). A discussion of U.S. national strategy and defense policies that impact security force assistance follows. This chapter then examines operations and programs related to security force assistance. Finally, it discusses the relationship of security force assistance and related operations and programs.

SECURITY FORCE ASSISTANCE AND THE STRATEGIC CONTEXT

1-1. *Security force assistance* is the unified action to generate, employ, and sustain local, host-nation or regional security forces in support of a legitimate authority (FM 3-07). Security force assistance (SFA) improves the capability and capacity of host-nation or regional security organization's security forces. These forces are collectively referred to as foreign security forces. ***Foreign security forces*** **are forces—including but not limited to military, paramilitary, police, and intelligence forces; border police, coast guard, and customs officials; and prison guards and correctional personnel—that provide security for a host nation and its relevant population or support a regional security organization's mission**.

1-2. SFA occurs within the framework of full spectrum operations (see FM 3-0). In most situations involving this assistance, there is relatively little weight on offensive and defensive operations from a U.S. perspective. However, when U.S. forces accompany foreign security forces (FSF) in combat, the weight of offensive and defense operations will change to address the situation and align with the foreign security force's efforts. SFA is not just a stability operation, although it is a key contributor to the primary stability tasks of establish civil security and establish civil control.

1-3. SFA can be conducted across the spectrum of conflict, from stable peace to general war. This assistance could focus on improving the security forces of a host nation that is currently under no immediate threat, on paramilitary forces to counter an insurgency, or on advising FSF in major combat operations against an external threat. Figure 1-1 illustrates how SFA spans the spectrum of conflict.

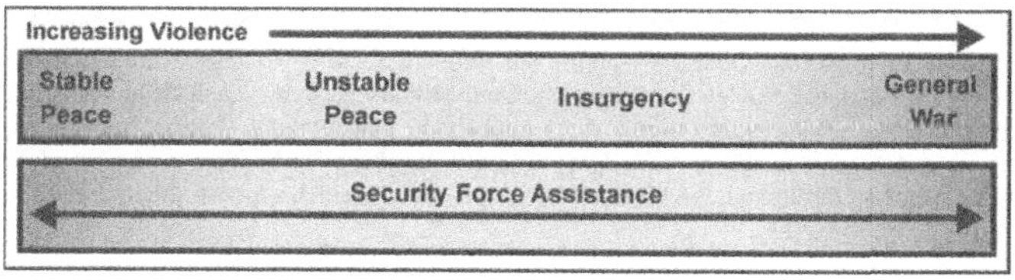

Figure 1-1. The spectrum of conflict and security force assistance

1-4. SFA can occur as part of any operational themes—peacetime limited engagement, limited intervention, peace operations, irregular warfare, and major combat operations. SFA often occurs during peacetime military engagement, peace operations, and irregular warfare. SFA involving paramilitary forces will often be limited to the irregular warfare operational theme, although paramilitary forces should eventually be absorbed, disbanded, or maintained in a long-term relationship with special operations forces.

1-5. While Army units are vital contributors, SFA is part of a comprehensive approach. This comprehensive approach should include close collaboration with military and civilian joint and multinational forces. The host-nation or regional security organization is the key actor within the comprehensive approach. It is vital that SFA is based on an objective, continuous assessment that examines the organization, training, equipment, rebuilding, and advising of the forces involved.

1-6. Law enforcement, military, intelligence, and border forces operate and cooperate within the security sector. U.S. forces understand how these units are intended to operate in the host-nation scheme, not necessarily using a U.S. model. U.S. forces should plan to help develop the respective capabilities so that these units can carry out their security functions.

1-7. Fully developed and institutionalized FSF plan, prepare, and execute at all the appropriate levels. Thus, U.S. forces must integrate SFA into the operations process at all levels. SFA should be designed to work seamlessly with the host-nation government at all levels, from ministries addressing the security sector to initial entry-level FSF training.

NATIONAL STRATEGY AND DEFENSE POLICIES

1-8. U.S. strategy and foreign policy affects all SFA. More importantly, SFA is critical to these policies as it is a primary tool for building partnership capacity with other nations. Key strategy documents for SFA include the National Security Strategy, the National Defense Strategy, the National Military Strategy, the Quadrennial Defense Review, and DODD 3000.05.

1-9. The National Security Strategy outlines the President's vision for providing enduring security for the American people in a volatile, uncertain, complex, and ambiguous strategic environment. The National Security Strategy of 2006 addresses stability operations within the broad engagement strategy for regional conflict. It supports SFA by "tailoring assistance and training of military forces to support civilian control of the military and military respect for human rights in a democratic society." This strategy also states that "[t]he world has found through bitter experience that success often depends on the early establishment of strong local institutions such as effective police forces and a functioning justice and penal system. This governance capacity is critical to establishing the rule of law and a free market economy, which provide long-term stability and prosperity."

1-10. The National Defense Strategy encompasses the Secretary of Defense's vision for the Department of Defense role in protecting the American people and U.S. interests. This strategy stems from the National Security Strategy. The 2008 National Defense Strategy highlights SFA. It states that "[o]ur strategy emphasizes building the capacities of a broad spectrum of partners as the basis for long-term security.... [l]ocal and regional conflicts in particular remain a serious and immediate problem...we will help build the internal capacities of countries at risk...[b]y helping others to police themselves and their regions, we will collectively address threats to the broader international system."

1-11. The National Military Strategy supports the National Security Strategy, National Defense Strategy, and Quadrennial Defense Review. The 2004 National Military Strategy states "[t]he Department will also work to strengthen alliances and partnerships by helping other nations increase their ability to defend themselves and protect common security interests." Working with other nations and strengthening partnerships appears throughout the National Military Strategy, which supports the importance of SFA.

1-12. The Quadrennial Defense Review is a congressionally mandated Department of Defense (DOD) review of strategy, programs, and resources. The 2006 Quadrennial Defense Review directed that joint ground forces "possess the ability to train, mentor and advise FSF and conduct counterinsurgency campaigns." This review also stated that conventional forces need to be capable of training, equipping, and advising local forces; deploying and engaging with partners; and supporting stability operations. Additionally, the review envisioned conventional ground forces continuing to perform more of the tasks traditionally performed by special operations forces. Conventional forces must understand foreign cultures and societies as well as be able to train, mentor, and advise FSF.

1-13. DODD 3000.05 provides guidance on stability operations. DODD 3000.05 establishes that stability operations are a core military mission and equal in importance to combat operations. DODD 3000.05 also reinforces the importance of SFA by stating "[t]he immediate goal often is to provide the local populace with security, restore essential services, and meet humanitarian needs. The long-term goal is to help develop host-nation capacity for securing essential services, a viable market economy, rule of law, democratic institutions, and a robust civil society."

RELATED OPERATIONS AND PROGRAMS

1-14. Several operations or programs directly or potentially relate to SFA. These operations and programs provide the context for SFA. They include security cooperation, security assistance, foreign internal defense, internal defense and development (IDAD), and security sector reform. For more information on operations related to SFA, see DOD 5105.38M, DOD security cooperation guidance, JP 3-07.1, FM 3-07, FM 3-05.137, and AR 11-31.

SECURITY COOPERATION

1-15. *Security cooperation* is all Department of Defense interactions with foreign defense establishments to build defense relationships that promote specific U.S. security interests, develop allied and friendly military capabilities for self-defense and multinational operations, and provide U.S. forces with peacetime and contingency access to a host nation (JP 3-07.1). Finally, security cooperation occurs across the spectrum of conflict and is not exclusively a peacetime activity. Security cooperation includes security assistance programs administered by the DOD as well as activities that enhance interoperability and the collective capability of combined forces (using Title 10 or exercise funding and authority). Security cooperation consists of a focused program of bilateral and multilateral defense activities conducted with foreign countries to serve mutual security interests and build defense partnerships. Security cooperation is governed by various sections of Title 10 and specific public laws addressing DOD interactions with other nations.

1-16. AR 11-31 establishes Department of the Army policy and prescribes responsibilities and procedures for the planning, integration, programming, budgeting, and execution of Army security cooperation activities. It is the policy of the U.S. Army to conduct security cooperation activities in compliance with higher-level guidance and in the execution of Army responsibilities under Title 10 and Title 22 which governs the transfer, exchange, conduct, and development of articles and services via a variety of U.S. Government programs.

1-17. Army security cooperation consists of official, cooperative, and noncombat interactions among any Army elements, Active Army, or Reserve Components. This includes any U.S. Government or nongovernment entity supporting the military and civilian joint and multinational organizations.

1-18. Commanders distinguish security cooperation from SFA based its emphasis in building relationships and capacities by using programmatic activities. Thus, security cooperation may support or be supported by SFA.

SECURITY ASSISTANCE

1-19. *Security assistance* is a group of programs authorized by the Foreign Assistance Act of 1961, as amended, and the Arms Export Control Act of 1976, as amended, or other related statutes by which the United States provides defense articles, military training, and other defense-related services by grant, loan, credit, or cash sales in furtherance of national policies and objectives (JP 3-57). Security assistance is a specific subset of security cooperation and may focus on external or internal threats.

1-20. DOD 5105.38-M describes the scope of security assistance programs in detail. Security assistance allows the transfer of military articles and services to friendly foreign governments. These transfers may be carried out via sales, grants, leases, or loans. If these transfers are essential to the security and economic well-being of allied governments and international organizations, they are equally vital to the security and economic well-being of the United States. U.S. forces can use security assistance to deliver defense weapon systems to foreign governments; to train international students; to advise other governments on improving internal defense capabilities; and to provide guidance and assistance in establishing infrastructures and economic bases for regional stability. Finally, security assistance cannot be conducted using Title 10 or exercise funds.

1-21. The military component of security assistance, implemented by DOD with policies established by the Department of State, has four principal components. They are international military education and training, foreign military sales, foreign military financing program, and peace operations. The foreign military financing program and international military education and training fall inside the military assistance budget process of the Department of State. Security assistance can also include funding peace operations.

International Military Education and Training

1-22. International military education and training contributes to internal and external security of a country by providing training to selected foreign militaries and related civilian personnel on a grant aid basis (see JP 3-07.1). These programs help to strengthen foreign militaries through training for the proper functioning of a civilian-controlled, apolitical, professional military. International military education and training serves as a foreign policy tool where the United States shapes doctrine; promotes self sufficiency in maintaining and operating U.S. acquired defense equipment; encourages the value of rule of law; and occasionally has a marked effect on the policies of the recipient governments. Foreign students, many of who occupy the middle and upper echelons of their country's military and political establishments, are taught U.S. doctrine and weapons systems employment resulting in greater cooperation and interoperability. All international military education and training activities are SFA.

Foreign Military Sales

1-23. Foreign military sales is a nonappropriated program through which foreign governments can purchase defense articles, services, and training from the United States (see JP 3-07.1). Eligible nations use this program to help build national security infrastructures. However, nations that require assistance are often unable to finance their needs. SFA supports foreign military sales when these sales include U.S. forces providing equipment and related training. It does not support foreign military sales when these sales provide equipment without related training.

Foreign Military Financing Program

1-24. The foreign military financing program provides funding to purchase defense articles and services, design and construction services, and training through foreign military sales or commercial channels (see JP 3-07.1). This program can assist nations with weak economies that would otherwise be unable to afford U.S. assistance. The foreign military financing program can fund foreign military sales. SFA and the foreign military financing program may be separate activities, although they can also be complementary.

Foreign Internal Defense

1-25. *Foreign internal defense* is the participation by civilian and military agencies of a government in any of the action programs taken by another government or other designated organization to free and protect its society from subversion, lawlessness, and insurgency (JP 3-05). U.S. foreign internal defense efforts involve all instruments of national power to support host-nation IDAD programs. The military instrument of foreign internal defense includes indirect support, direct support (not involving U.S. combat operations), and combat operations. Foreign internal defense can occur across the spectrum of conflict and within any operational theme. Foreign internal defense is often a part of irregular warfare or peacetime military engagement operational theme. SFA may support the military instrument of foreign internal defense. JP 3-07.1 discusses foreign internal defense in detail.

Note: It is a national policy decision for U.S. forces to accompany FSF on combat operations.

Military Instrument of Foreign Internal Defense

1-26. The military instrument of foreign internal defense supports the other instruments of national power through activities ranging from peacetime military engagement to major combat operations. Although foreign internal defense may include capacity building across the host nation, U.S. forces focus on combating internal threats to help the host nation maintain legitimacy and influence over the population. As such, foreign internal defense may be part of irregular warfare.

1-27. SFA and foreign internal defense support the host nation's IDAD. Consequently, forces conducting SFA must have proper mindset—they are part of a developmental and supporting effort. This applies even if SFA is part of working with a potential adversary or an existing enemy. SFA and foreign internal defense focus on working by, with, and through FSF and not conducting unilateral efforts to achieve objectives.

Indirect Support

1-28. Indirect support efforts are foreign internal defense operations that emphasize the principle of host-nation self-sufficiency. Indirect support focuses on building strong national infrastructures through economic and military capabilities that contribute to self-sufficiency. For example, indirect support could include U.S. assistance in improving a host-nation's intelligence infrastructure. Contributions from U.S. forces to indirect support come from security cooperation guidance through security assistance. They are supplemented by multinational exercises, exchange programs, and selected joint exercises. Indirect support that is focused on assisting security forces is SFA.

1-29. Although they indirectly assist those involved, military exchange programs are not SFA. Military exchange programs foster mutual understanding between forces; familiarize each force with the organization, administration, and operations of the other; and enhance cultural awareness. These exchange programs often have long-term implications for strengthening democratic ideals and respect for human rights among supported governments. However, such exchange programs (funded with Title 10 monies) do not themselves become vehicles for security assistance training, as they do not intend to assist security forces but to foster mutual understanding and enhance cultural awareness.

1-30. Joint and multinational exercises strengthen U.S. and host-nation relations and interoperability of forces, but they are not SFA. They complement security assistance and civil-military operations by validating host-nation needs and capabilities and by providing a vehicle for the conduct of humanitarian and civic assistance programs (see JP 3-07.1). Strict legal restrictions clarify the type of support that U.S. forces can provide and on the monetary limits of such support. These efforts are not SFA as they do not intend to directly assist security forces; instead, they intend to improve relations and interoperability. See appendix B.

Direct Support

1-31. Direct support does not involve combat operations but includes the use of U.S. forces providing direct assistance to the host-nation civilians or military. Direct support often focuses on civil-military operations (primarily, providing services to local civilians), psychological operations, intelligence and communications sharing, and logistic support. Sometimes direct support will be SFA, such as psychological operations, military training support, logistic support, and intelligence and communications sharing. SFA is not a civil affairs activity such as foreign humanitarian assistance, humanitarian and civic assistance, or military civic action.

1-32. Direct support differs from security assistance in that direct support is joint- or Service-funded. Direct support does not usually involve transferring arms and equipment and does not often include training local military forces. Direct support operations are normally conducted when the host nation has not attained or regained self-sufficiency and when it still faces social, economic, or military threats beyond its capability.

Combat Operations

1-33. The introduction of U.S. combat forces into foreign internal defense operations requires a Presidential decision. These operations are a temporary solution until host-nation forces can stabilize the situation and provide security for civilians. In all cases, U.S. combat operations support the host-nation IDAD program and remain strategically defensive in nature. When U.S. forces are advising, mentoring, partnering, or augmenting host-nation forces in foreign internal defense combat operations, this is SFA. When Army units conduct independent combat operations alongside FSF, this is not SFA.

1-34. U.S. tactical participation in host-nation internal conflicts requires judicious and prudent rules of engagement and guidelines for applying force. Inappropriate destruction and violence attributed to U.S. forces may easily reduce the legitimacy and sovereignty of the supported government. Adversaries may also use such incidents to fuel anti-American sentiments and assist their cause.

Security Relationships

1-35. A complex relationship exists among security cooperation, security assistance, and the military instrument of foreign internal defense. The left side of figure 1-2 depicts this relationship, including how aspects of foreign internal defense and security assistance overlap. The right side of figure 1-2 illustrates how foreign internal defense focuses on internal threats to a host nation and how security assistance focuses on external threats. The column depicts how security (military, intelligence, and civilian), economic, and governance are considerations common to both foreign internal defense and security assistance.

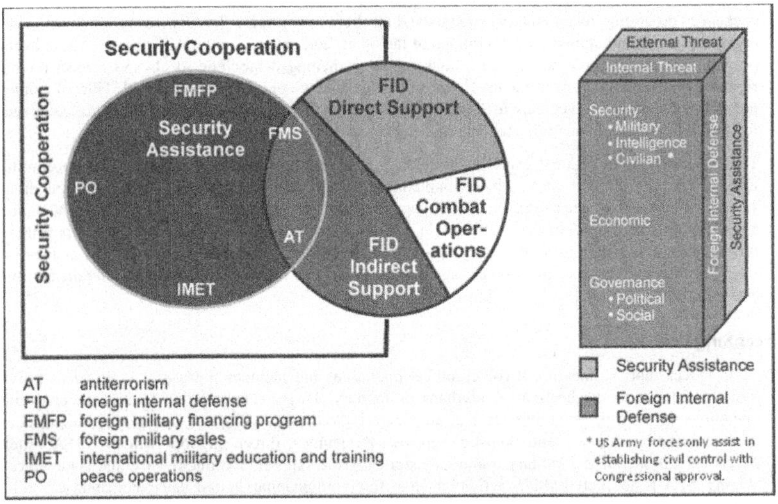

Figure 1-2. Security cooperation, security assistance, and foreign internal defense

Internal Defense and Development

1-36. *Internal defense and development* is the full range of measures taken by a nation to promote its growth and to protect itself from subversion, lawlessness, and insurgency. It focuses on building viable institutions (political, economic, social, and military) that respond to the needs of society (JP 3-07.1). IDAD aims to forestall or defeat the threat and to correct core grievances that prompt violence. Ideally it is a preemptive strategy. However, if an insurgency or other threat develops, it becomes an active strategy to combat that threat. To support the host nation effectively, U.S. forces, especially planners, consider the host-nation's IDAD strategy.

1-37. IDAD focuses on building viable institutions that respond to the needs of society. IDAD is a U.S. doctrinal term; it is not used universally by all countries. Internal defense and development blends four interdependent functions to prevent or counter internal threats: balanced development, security, neutralization, and mobilization. Finally, IDAD involves a cyclic interaction of execution, assessment, and adaptation. JP 3-07.1 discusses IDAD and its functions in detail.

Security Sector Reform

1-38. *Security sector reform* is the set of policies, plans, programs, and activities that a government undertakes to improve the way it provides safety, security, and justice (FM 3-07). Security sector reform aims to provide an effective and legitimate public service that is transparent, accountable to civil authority, and responsive to the needs of the public. It may include integrated activities to support defense and armed forces reform; civilian management and oversight; justice, police, corrections, and intelligence reform; national security planning and strategy support; border management; disarmament, demobilization, and reintegration; and concurrent reduction of armed violence. SFA is a vital component of security sector reform when security sector reform includes U.S. assistance to FSF. See chapter 6 of FM 3-07 for an additional discussion of security sector reform.

Security Force Assistance Relationship

1-39. Paragraph 1-35 explained and figure 1-2 supported the complex relationship among security cooperation, security assistance, and the military instrument of foreign internal defense. Figure 1-3 explains the relationship of SFA to security cooperation, security assistance, and the military instrument of foreign internal defense. SFA supports the military instrument of foreign internal defense, much of security assistance efforts, and some security cooperation efforts. The right side of figure 1-3 also shows that SFA addresses security issues, but not economic or governance issues.

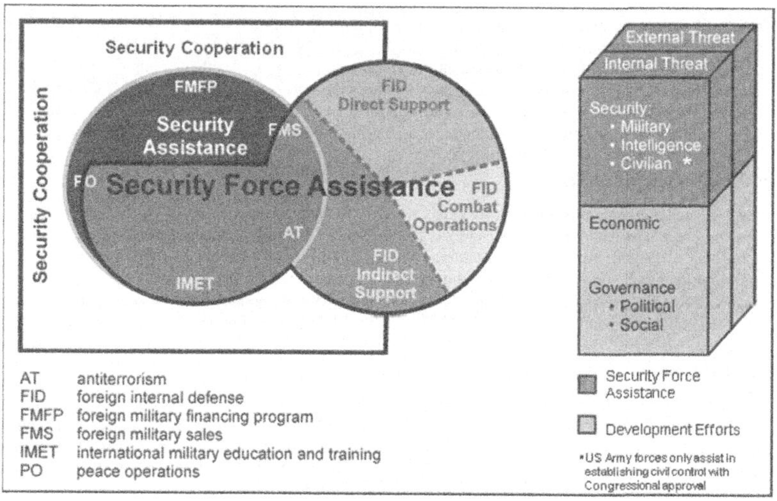

Figure 1-3. Relationship of security force assistance with security cooperation, security assistance, and foreign internal defense

Country Team

1-40. The country team will play a vital role with respect to SFA. The country team is the senior in-country U.S. coordinating and supervising body headed by the U.S. chief of mission. The chief of mission leads the country team and integrates U.S. efforts in support of the host nation. The chief of mission is the highest U.S. civil authority in a foreign country. As the senior U.S. Government official permanently assigned in the host nation, the chief of mission is responsible to the President for policy oversight of all U.S. Government programs. The Foreign Service Act assigns the chief of mission to a foreign country with responsibility for the direction, coordination, and supervision of all government executive branch employees in that country except for Service members and employees under the command of a U.S. area military commander.

1-41. The country team is composed of the senior member of each represented department or agency. As permanently established interagency organizations, country teams represent a priceless resource. They often provide deep reservoirs of local knowledge and interaction with the host-nation government and population. See FM 3-07 and JP 3-07.4 for more information on the country team.

Theater Command Structures

1-42. SFA is inherently multinational. Nations do not relinquish their national interests by participating in multinational operations. As in any multinational operation, commanders conducting SFA prepare to address issues related to legality, mission mandate, and prudence early in the planning process. In multinational operations, consensus often stems from compromise.

1-43. Political considerations heavily influence the ultimate shape of the command structure in which U.S. forces will conduct SFA. However, participating nations should strive to achieve unity of command for the operation to the maximum extent possible. All participants must understand missions, tasks, responsibilities, and authorities. While well defined in U.S. doctrine, command relationships are not necessarily part of the doctrinal lexicon of nations with which the United States may operate in coalition or combined operations.

1-44. Command and control relationships of higher echelons take precedence when conducting SFA. As a result, the brigade combat team (BCT) may be a supported command from a U.S. perspective as well as acting as a supporting command for FSF in the BCT's area of operations. For example, a BCT may support a joint interagency task force conducting international police training. Another example is a joint command supporting a BCT with joint fires; intelligence, surveillance, and reconnaissance; or other capabilities while the BCT supports FSF in training. The basic structures for multinational operations fall into one of three types: lead nation, integrated, or parallel command. Alliance and coalition commands use these structures.

Lead Nation Command Structure

1-45. A lead nation command structure exists when all member nations place their forces under the control of one nation. The lead nation command has a dominant lead nation command and staff arrangement with subordinate elements retaining strict national integrity. This command may also have an integrated staff and multinational subordinate forces. Integrating the staff allows the commander to draw upon the expertise of allied or coalition partners in areas where the lead nation may have less experience. This is the most desirable command structure for SFA.

Integrated Command Structure

1-46. Multinational commands organized under an integrated command structure provide unity of effort in a multinational setting. An integrated, combined command has several distinguishing features. Besides a designated single commander, the staff consists of representatives from all member nations. Subordinate commands and staffs are integrated into the lowest echelon necessary. Although effective for SFA, the integrated command structure works less effectively than the lead nation command structure.

Parallel Command Structure

1-47. Under a parallel command structure, no single force commander is designated. The coalition leadership develops a means for coordination among the participants to attain unity of effort. They can accomplish this by using coordination centers. Nonetheless, because of the absence of a single commander, the use of a parallel command structure should be avoided if at all possible. This is the least desirable command structure for SFA.

Alliance Command Structures

1-48. In combined commands, national political objectives are addressed and generally subsumed within multinational force objectives at the alliance treaty level. Combined command relationships often reflect either an integrated command structure or a lead nation command structure. In combined operations such as SFA, these structures should be used to the maximum practical extent. Combined command and force structures often mirror the degree of allied member participation. Senior military officers from member nations often lead subordinate commands. Effective operations within an alliance, including SFA, require senior political and military authorities to agree on the type of command relationships that will govern the operations of the forces. Potential political sensitivities associated with SFA can impact command relationships and operating procedures.

Coalition Command Structures

1-49. Coalitions often form in response to unforeseen crises that occur outside the area of an established alliance, which is often the case for SFA. Coalition command relationships routinely evolve as a coalition develops. Coalitions often have one of two basic structures: lead nation or parallel. For SFA, all participating coalition members should augment the host-nation headquarters staff with representatives. These representatives might include designated deputies or assistant commanders, planners, and logisticians. Such staffing provides the coalition commander with representative leadership and accessible expertise on the capabilities of the coalition members as well as facilitates the planning process. This integration of the multinational command elements into the coalition planning process should occur as early as practical.

Integration of Conventional and Special Operations Forces

1-50. Many criteria determine when conventional forces, special operations forces, or a combination are appropriate to conduct SFA. Both force levels and force characteristics suggest optimal, acceptable, and undesirable force package options in planning and resourcing SFA. Options for the deployment of a modular brigade augmented for SFA, a select number of conventional military transition teams, or special operations forces depend on conditions of the operational environment, priorities of the IDAD strategy, overall U.S. national policy, and forces available.

1-51. Rarely will U.S. forces conducting SFA be homogenous in terms of conventional or special operations forces, but rather contain both. FM 3-0 and other Army doctrine define command and control relationships and support relationships when both forces operate in the same area. Recent operational experience has shown that, in improving the effectiveness of FSF, the supported commander conducting SFA may be either a conventional force or special operations force commander. Subordinate commanders and staffs of both conventional and special operations forces may act in the supporting role to the U.S. ambassador. Typically this is done through the security assistance officer or the defense attaché at the American Embassy.

1-52. If U.S. forces are to train a large number of FSF in a short time, then the requirement may exceed the capacity of available special operations forces. Large SFA missions occur at the geographic command level or higher headquarters. Planners at these levels determine the number of conventional and special operations forces as part of their mission analysis aimed at training enough FSF in the given time frame. To best prepare for a large SFA, U.S. forces might deploy one or more augmented modular brigades supported by Army special operations forces. These forces conduct complementary foreign internal defense tasks under Title 22 authorities. Another option involves using special operations forces training FSF over several years while using a broad mix of funding and authorities. Another option uses a modular brigade to support a select number of company- or battalion-sized foreign units.

1-53. A force analysis of how best to integrate conventional and special operations forces to accomplish SFA considers host-nation and U.S. national policies. Skill sets among conventional and special operations forces trainers, advisors, and mentors avoid universally excluding or including final, integrated force packages for SFA. Planners leverage skill sets by integrating and coordinating conventional and special operations units with assigned tasks within SFA. Force analysis of the FSF to be trained often suggests a two-phase approach to training. This analysis also suggests a transition to the FSF by either conventional or special operations forces using basic and advanced training. The type of training depends on the nature of the foreign security force being trained. Additionally, integrating the advisor may be a phased process using either conventional or special operations forces as the initial advisor. This method lends itself to the classic crawl-walk-run approach to assisting FSF.

Chapter 2
Framework

This chapter discusses a doctrinal framework for the conduct of security force assistance. This security force assistance framework is composed of the mindset required of units and Soldiers, imperatives for success, inherent tasks and activities, and the three types of security force assistance.

MINDSET

2-1. Conducting successful security force assistance (SFA) requires a specific mindset. This mindset focuses on working by, with, and through foreign security forces (FSF) to support the host nation's internal defense and development (which includes local security requirements) or regional organization's charter. Soldiers conducting SFA must also understand that legitimacy is vital. The relevant population must perceive FSF as legitimate for long-term success. Those conducting SFA must understand that the military instrument of national power is only one part of a comprehensive approach. The imperatives of SFA provide the foundation for proper mindset.

***Note*:** Host nation includes partner organization.

2-2. SFA focuses on how FSF organizations, institutions, systems, capabilities, and limitations—not U.S. organizations, systems, procedures, and equipment—can be enabled to address the host-nation or partner organization problems. Local forces have advantages over outsiders. They inherently understand the local culture and behavior that outsiders simply lack. To tap into those advantages, advisors must resist blatant military solutions. To overcome the temptation to do what they know and do best, whether relevant or not to the situation, advisors must accept that they are bound by unique situations.

IMPERATIVES

2-3. The imperatives of SFA come from the historical record and recent experience. These imperatives do not replace the principles of war or the principles of joint operations. Rather, they provide focus on how to successfully conduct SFA. The six imperatives apply to SFA at every level of war, for any echelon, and for any Soldier.

2-4. Since there is a close relationship between SFA and Army special forces conducting foreign internal defense, SFA planners should also consider the twelve Army special operations forces imperatives, especially at the tactical level (see appendix A).

UNDERSTAND THE OPERATIONAL ENVIRONMENT

2-5. An in-depth understanding of the operational environment—including available FSF, opposing threats, and civil considerations—is critical to planning and conducting effective SFA. Units and Soldiers conducting SFA must clearly understand the theater, population, and FSF with which they are working, especially FSF capabilities. Diplomatic, informational, military, economic, sociological, psychological, and geographic research and understanding are essential prerequisites for successful SFA. Tactically, successful SFA requires identifying the friendly and hostile decisionmakers, their objectives and strategies, and the ways they interact. Further, the conditions of the operational environment can rapidly change, and those conducting SFA must anticipate these changes and exploit potential and possibly fleeting opportunities.

PROVIDE EFFECTIVE LEADERSHIP

2-6. Leadership, a critical aspect of any application of combat power, proves important in the dynamic and complex environments associated with SFA. The operational environment in which SFA occurs places a premium on effective leadership at all levels, from the most junior to the most senior general officer and agency director. Leading FSF or a combined group of U.S. and foreign security forces is inherently more challenging due to differences in culture, language, training, and other aspects. Leaders focus on transitions; their decisions move the foreign security force toward independent operations.

BUILD LEGITIMACY

2-7. Ultimately, SFA aims to develop security forces that contribute to the legitimate governance of the local populace. Significant policy and legal considerations may apply to SFA activities. Legitimacy is the most crucial factor in developing and maintaining internal and international support. The United States cannot sustain long-term SFA without legitimacy. Legitimacy is a concept that goes beyond a strict legal definition; it includes the moral and political legitimacy of a host-nation government or partner organization. Local civilians and the international community determine the government's legitimacy based on collective perception of the credibility of its cause and methods.

MANAGE INFORMATION

2-8. Successful SFA disseminates timely and protected relevant information, integrates it during planning, and leverages it appropriately during execution. Effective and efficient information management supports decisionmaking throughout capacity building. Managing information encompasses the collection, analysis, management, application, and preparation of information. Maintaining continuity between rotating SFA elements is critical. For long-term SFA efforts, FSF expect more professional abilities from U.S. advisors. To meet expectations and success in general, units conducting SFA establish and integrate lessons learned. Capturing and incorporating evolving tactics, techniques, and procedures and aspects of the operational environment is vital; with each successive rotation, planners strive to maintain continuity and momentum. Finally, planners synchronize effective information engagement with the entire mission. See appendix C. SFA operations inherently affect the dynamic operational environment.

ENSURE UNITY OF EFFORT

2-9. SFA often includes many actors, making unity of effort essential for success. SFA will include U.S. and foreign security forces, including conventional forces, special forces, or a combination. Other civilian and military joint and military organizations are often involved in SFA. Planners integrate them into one cohesive effort.

SUSTAIN THE EFFORT

2-10. Sustainability consists of two major components: the ability to sustain SFA effort throughout the operation and the ability of the FSF to sustain their operations independently. While each situation will vary, Army personnel conducting SFA must avoid assisting FSF in techniques and procedures beyond the FSF's capability to sustain. U.S. tactics, techniques, and procedures must be modified to fit the culture, educational level, and technological capability of the FSF. Those involved in SFA must recognize the need for programs that are durable, consistent, and sustainable by both the U.S. and FSF. They must not begin programs that are beyond the economic, technological, or cultural capabilities of the host nation to maintain without U.S. assistance. Such efforts are counterproductive.

TASKS

2-11. SFA aims to establish conditions that support the partner's end state, which includes legitimate, credible, competent, capable, committed, and confident security forces. This requires a force capable of securing borders, protecting the population, holding individuals accountable for criminal activities, regulating the behavior of individuals or groups that pose a security risk, and setting conditions in the operational area that enable the success of other actors.

2-12. This section addresses the six SFA tasks: organize, train, equip, rebuild and build, and advise and assist. When supporting the development of FSF, commanders understand their role with the host-nation or regional security organization. These tasks facilitate SFA planners to assess and allocate resources based on conditions.

ORGANIZE

2-13. Organize is a SFA task that encompasses all measures taken to assist FSF in improving its organizational structure, processes, institutions, and infrastructure. U.S. forces must understand the existing security organizations of FSF to better assist them. Subsequently, SFA personnel may help the host nation organize its security forces to meet the needs of its security environment.

2-14. Organizing a foreign security force depends on the host nation's social and economic conditions, cultural and historical factors, and security threats. SFA aims to create an efficient organization with a command, intelligence, logistic, and operations structure viable for the host nation. Conventional forces with limited special purpose teams (such as explosive ordnance disposal) are preferred. Doctrine or standing operating procedures should apply across the force, as should unit structures. The organization must facilitate the collection, processing, and dissemination of intelligence across and throughout all security forces. As the foreign security force strengthens, U.S. leaders and trainers should expect more independent organizational decisions.

2-15. The host nation determines the structure of its military forces, to include approving all organizational designs. These may include changing the numbers of forces, types of units, and internal organizational designs.

2-16. Organization should address all FSF, from the ministerial level to the patrolling police officer and soldier. Building a competent FSF infrastructure—including civilian information systems—is critical for success. Commanders work with ministries responsible for national and internal security, including the ministry of national defense, the interior ministry, and the justice ministry. Commanders assess strengths and weaknesses of the ministerial organization as well as training requirements of their employees. The joint and multinational advisor team at the ministry level helps the host nation develop a procurement and management system to meet its requirements.

2-17. Organizing a foreign security force requires resolving issues related to—

- Recruiting.
- Promotion screening and selection.
- Pay and benefits.
- Leader recruiting and selection.
- Personnel accountability.
- Demobilization of security force personnel.

Recruiting

2-18. Recruiting is critical when establishing FSF. The recruiting program assimilates local culture and themes that resonate with the populace. The program ensures that FSF have members from all major demographic groups. Joint and multinational partners should encourage and support efforts to recruit from minority populations. Mobile recruiting can target specific areas, ethnic groups, or tribes. Forces should encourage moderate groups and factions within hostile or potentially hostile ethnic groups to join the foreign security force. Often host-nation governments resist recruiting disaffected ethnic groups. However, even moderate success in recruiting from these groups provides enormous payoffs. It builds the security forces' legitimacy and often quiets legitimate fears of such groups regarding their relationship to the host-nation government. Effectively disarming, demobilizing, and reintegrating former enemies or other armed groups should be part of the overall plan. These programs should be mentioned in recruiting efforts.

2-19. A proper recruiting program specifies appropriate behavioral, physical, and moral standards. Ideally, recruits are centrally screened and inducted. Recruiting centers reside in safe areas safe. All recruits undergo a basic security check and are vetted against lists of suspected enemies. Foreign agencies and personnel perform this screening. Membership in illegal organizations needs to be carefully monitored. Past membership need not preclude a person from joining the foreign security force; however, any ongoing relationship with an illegal organization requires constant monitoring. FSF personnel limits the members of an illegal unit, tribal militia, or other militant faction to a single military or police unit.

Promotion Screening and Selection

2-20. Selection for promotion must stem from proven performance and aptitude for increased responsibility. Objective evaluations ensure promotion is by merit and not through influence or family ties. One method of selection identifies the most competent performers, trains them, and recommends them for promotion. The second method identifies those with social or professional status within the training group, then trains and recommends them for promotion. The first method may lead to more competent leaders but could be resisted for cultural reasons. The second method ensures the new leader will be accepted culturally but may sacrifice competence. The most effective solution comes from combining the two methods.

Pay and Benefits

2-21. Appropriate compensation precludes a culture of corruption in FSF. Spending the money needed for adequate wages and producing quality security forces is less costly than ending up with corrupt and abusive forces that alienate the populace. Paying the police adequately is especially important; their duties and contact with civilians often expose them to opportunities for corruption.

2-22. Payroll procedures and systems are also vitally important. They must be transparent and accountable, so FSF members receive their full pay and entitlements. Centralized administration of compensation through secure, automated systems minimizes the risk for corruption and manipulation; however, the supporting infrastructure will have to be available or developed over time.

2-23. Effective FSF can help improve host-nation social and economic development through the benefits each member receives. Every recruit should receive a basic education, job training, and morals and values inculcation.

Leader Recruiting and Selection

2-24. Officer candidate standards should be high. Candidates should have good health and pass an academic test with higher standards than the test for enlisted recruits. Recruiters carefully vet officer candidates to ensure that they do not have close ties to any radical or enemy organization.

2-25. Noncommissioned officers should be selected from the best enlisted FSF members. Objective standards, including proficiency tests, should be established and enforced to ensure that promotion to their ranks comes from merit. Many armies lack a professional noncommissioned officer corps; establishing one for a foreign security force may prove difficult. In the meantime, adjustments will have to be made, placing more responsibility on commissioned officers.

Personnel Accountability

2-26. Host-nation leaders carefully track and account for FSF. Proper personnel accountability reduces corruption, particularly in countries with manual banking systems where soldiers are paid in cash. In addition, many personnel failing to report for duty can indicate possible attacks, low unit morale, or enemy and militia influences on the foreign security force.

Demobilization of Foreign Security Force Personnel

2-27. Host nations develop programs to keep a class of impoverished and disgruntled former officers and soldiers from forming. As the foreign security force mature, officers who perform poorly or fail to meet the standards will need to be removed. Providing some form of government-provided education grants or low-interest business loans enables discharged personnel to earn a living outside the military. All soldiers who serve for several years and are then removed should receive a lump-sum payment or pension to ease their transition to civilian life. These programs should not apply to those guilty of major human rights abuses or corruption. Transition or demobilization planning should start as soon as commanders anticipate the need (it may not be required in all cases). Similar programs may be required when demobilizing nongovernment militias.

TRAIN

2-28. Train is a SFA task to assist FSF by developing programs and institutions to train and educate. These efforts must fit the nature and requirements of their security environment. The imperatives of Army special operations forces for facilitating a comprehensive approach and engaging the threat discriminately apply to the train task (see appendix A).

Training the U.S. Trainers

2-29. Soldiers assigned training missions should receive training on the requirements of developing FSF. The training should emphasize the host nation's cultural background, introduce its language, and provide cultural tips for developing a good rapport. Training should also include protection for troops working with FSF. U.S. trainees must become familiar with FSF organization and equipment, especially weapons not in the U.S. inventory. This training must emphasize the following:

- Sustaining training and reinforcing individual and team skills.
- Using the smallest possible student-to-instructor ratio.
- Developing host-nation trainers.
- Training to standards—not to time.
- Providing immediate feedback; using after action reviews.
- Respecting the culture but learning to distinguish between cultural practices and excuses.
- Learning the language.
- Working with interpreters.

2-30. U.S. forces should show respect for local religions and traditions. They should willingly accept many aspects of the local and national culture, including food (if sanitation standards permit). U.S. forces must stress that they do not intend to undermine or change the local religion or traditions. However, they have to strive to reduce how dysfunctional social practices affect the foreign security force's ability to conduct operations. U.S. trainers and advisors must have enough awareness to identify and stop inappropriate behavior, or at least report it to the appropriate chains of command.

Establishing Training Standards

2-31. Training in SFA involves many of the individual and collective skills performed in conventional military operations. All levels of training for all components should include values training. Metrics for evaluating units should include subjective measures, such as loyalty to the host-nation government, as well as competence in military tasks. Soldiers know how to evaluate military training. However, the acceptance of values, such as ethnic equality or the rejection of corruption, may be a better measure of training effectiveness. Gauging this acceptance is far more difficult than evaluating task performance. While the operational environment varies widely, FSF and trainers can still establish clear measures to evaluate the training of individuals, leaders, and units.

2-32. Effective training programs require clear, detailed individual, leader, and unit performance standards. These standards take into account cultural factors that directly affect the ability of the individual or unit to operate. For example, training a staff or unit to conduct effective operations requires more time in countries where the average soldier is illiterate. Building a security force from the ground up takes far more time than creating one around a trained cadre. Thus, using existing military personnel to form units and cadres for units often proves better than creating novice security forces.

2-33. Often, poorly trained leaders and units commit more human rights violations than well-trained, well-led units. Leaders and units unprepared for the pressure of active operations tend to use indiscriminate force, target civilians, and abuse prisoners. These actions can threaten the popular support and government legitimacy essential for SFA success. Badly disciplined and poorly led FSF can facilitate insurgent recruiting and propaganda efforts.

2-34. Setting realistic measures for FSF and following through on training plans consume time. The pressure is strong to find training shortcuts, employ quick fixes, or train personnel on the job. Trainers should resist such approaches. Such approaches often create more problems than they solve. However, trainers should also avoid the temptation to create long, complex training programs based on unrealistic standards. Effective programs account for the host nation's culture, resources, and short-term security needs. No firm rules exist on how long particular training programs should take. Trainers can use existing and historical training programs to determine how long training should take. To a certain extent, the enemy threat dictates the length of training. As security improves, training programs can expand to facilitate achievement of the long-term end state.

Training Foreign Security Forces

2-35. Members of FSF develop through a systematic training program (individual training and education as well as collective training). The program builds basic skills, then teaches them to work as a team, and finally allows them to function as a unit. Basic military, intelligence, or law enforcement training focuses first on basic skills such as legal considerations, first aid, marksmanship, and fire discipline. Leaders have training in tactics, including patrolling, urban operations, and legal evidence collection. Everyone must master rules of engagement and the law of armed conflict. FSF units should train to standard for conducting the major operations they will face.

Training Leaders

2-36. The effectiveness of FSF directly relates to the quality of their leadership. Building effective leaders requires a comprehensive program of officer, staff, and specialized training. The ultimate success of any SFA effort depends on creating viable FSF leaders able to carry on the fight at all levels and build their nation on their own.

2-37. The standards of leader training reinforce different levels of authority within the foreign security force. Clearly established responsibilities for commissioned and noncommissioned officers specify what is expected of recruits and leaders. To ensure civilian control, subordinate relationships to civilian authorities also are reinforced. In addition, training should establish team dynamics. In some cultures, security forces may need training to understand the vital role of members not in primary leadership positions. For example, noncommissioned officers may be a new or different concept for some FSF.

Employing Newly Trained Forces

2-38. SFA must build the morale and confidence of FSF. Committing poorly trained and badly led forces results in high casualties and invites tactical defeats. While defeat in a small operation may have little strategic consequence in a conventional war, often a small tactical defeat of FSF has serious strategic consequences. If the foreign security force fails, the local populace may begin to lose confidence in the host-nation government's ability to protect them. FSF must prepare for operations so they have every possible advantage. As much as possible, FSF should begin with simpler missions. As their confidence and competence grows, these forces can assume more complex assignments. Collaborating with joint or multinational units can help new forces to adjust to combat stress.

2-39. Newly trained units should enter their first combat operation in support of more experienced foreign units. Operational performance of such inexperienced organizations should be carefully monitored and evaluated to correct weaknesses quickly. The employment plan for FSF should allow enough time for additional training after each operation. Gradually introducing units into combat allows the command to identify poor leaders for retraining or other actions. Competent leaders are also identified and given greater authority and responsibility.

EQUIP

2-40. Equip is a SFA task encompassing all efforts to assess and assist FSF with the procurement, fielding, and sustainment of equipment. All equipment must fit the nature of the operational environment. The SFA principle of ensuring long-term sustainment is a vital consideration for the equip task.

2-41. The strategic plan for security force development should outline equipment requirements. Appropriate equipment is affordable and suitable against the threat. Forces must be able to train on the equipment. Interoperability may be desired in some cases. A central consideration includes the host nation's long-term ability to support and maintain the equipment.

2-42. The initial FSF development plan should use a comprehensive approach with local, regional, national, and other appropriate actors. Fundamentally, it should support the internal defense and development strategy. It should include phases with objectives for FSF to meet over three to four years. Since potential enemies adapt rapidly and situations change, commanders must continually assess the direction and progress of developing FSF.

2-43. The requirement to provide equipment may be as simple as assisting with maintenance of existing formations or as extensive as providing everything from shoes to communications and investigation kits. If the enemy uses heavy machine guns and rocket-propelled grenades, then FSF need comparable or better equipment. This especially applies to police forces, which are often lightly armed and vulnerable to well-armed enemies.

2-44. Primary considerations should include maintainability, ease of operation, and long-term sustainment costs. Few developing nations can support highly complex equipment. In most operations, having many versatile vehicles that require simple maintenance is often better than having a few highly capable armored vehicles or combat systems that require extensive maintenance. Effective maintenance systems for FSF often begin with major maintenance performed by contractors. The program then progresses to arrangements with U.S. forces as they train foreign personnel to conduct the support mission.

2-45. Sources of materials for FSF include U.S. foreign military sales (not technically SFA), multinational or third-nation resale of property, contracts with internal suppliers, or purchases on the international market. The organizations responsible for equipping FSF require flexibility to obtain equipment that meets their needs for quality, timeliness, and cost. As part of their training, FSF also need to learn property accountability to reduce corruption and ensure proper equipment usage. FSF provide equipment the same level of control and protection that U.S. forces provide for similar equipment. (See AR 12-1 and DOD 5105.38M.)

REBUILD AND BUILD

2-46. Rebuild and build is a SFA task to assess, rebuild, and build the existing capabilities and capacities of FSF and their supporting infrastructure. This task requires an in-depth analysis of the capability, capacity, and structures required to meet the desired end state and operational environment. Some FSF may require assistance in building and rebuilding, while other FSF may only need assistance in building.

2-47. FSF need infrastructure support. People need buildings for storage, training, and shelter. Often requirements include barracks, ranges, motor pools, and other military facilities. Construction takes time; the host nation needs to invest early in such facilities if they are to be available when needed. Forces must plan to protect any infrastructures, including headquarters facilities, since these building are attractive targets for insurgents. (See FM 3-34.400 for information on hardening measures to increase infrastructure survivability and improve protection.)

2-48. During some operations, such as counterinsurgency, foreign military and police forces often operate from local bases. Building training centers and unit garrisons requires a long-term force-basing plan. If possible, garrisons should include housing for the host-nation soldiers and their families; government-provided healthcare for the families; and other attractive benefits.

2-49. The host nation may need to make large investments in time and resources to restore or create the nationwide infrastructure necessary to effectively command and control FSF. The host nation will build functional regional and national headquarters and ministries as well as local bases and police stations.

Advise and Assist

2-50. Advise and assist is a SFA task in which U.S. personnel work with FSF to improve their capability and capacity. Advising establishes a personal and a professional relationship where trust and confidence define how well the advisor will be able to influence the foreign security force. Assisting is providing the required supporting or sustaining capabilities so FSF can meet objectives and the end state. The level of advice and assistance is based on conditions and should continue until FSF can establish required systems or until conditions no longer require it. Leaders cannot permit FSF to fail critically at a point that would undermine the overall effort. See chapter 6 for a discussion on advisors.

ACTIVITIES

2-51. SFA is inherently a developmental effort. Success is measured by the increase in the foreign security force's capability, capacity, competency, commitment, and confidence in areas in which U.S. forces are assisting them and as a whole. SFA activities include plan and resource, generate, employ, transition, and sustain. Figure 2-1 depicts the SFA activities. These activities blend over time based on the capability and capacity of FSF.

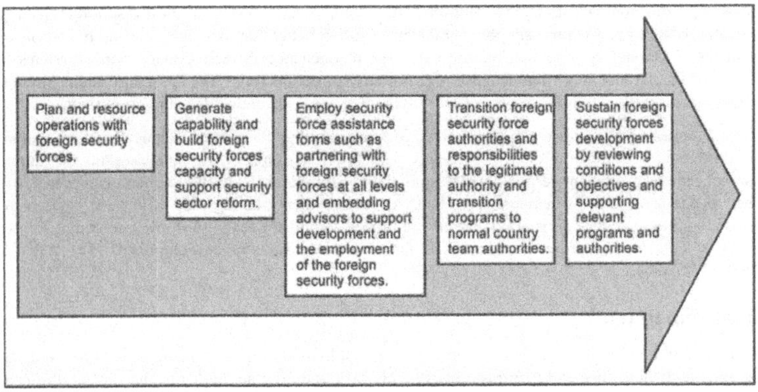

Figure 2-1. Security force assistance activities

Plan and Resource

2-52. The plan and resource SFA activity begins as commanders understand the operational environment and determine the requirements of FSF. It also ensures that the United States provides SFA that achieves the objectives and end state of both the host nation and United States. The host nation and United States should then analyze the resource requirements and efforts so that developing FSF have sufficient and appropriate resources. Chapter 4 provides additional detail on planning for facilities and infrastructure.

Generate

2-53. The generate SFA activity includes generation of FSF, including the forces, leaders, and warfighting function capabilities (movement and maneuver, intelligence, fires, sustainment, command and control, and protection) based on the desired end state.

Employ

2-54. The employ SFA activity involves FSF transitioning from force generation to mission employment. It does not rely on the maturity of the force or supporting institutions as a whole but is focused on the foreign element in question. Conditions determine when to use an element of FSF. Leaders of U.S. and foreign security forces assess the risk associated with employment and mitigate that risk as much as possible.

Transition

2-55. The transition SFA activity defines the transition between two security forces, when applicable. This could be a transition of authority between U.S. forces to a new foreign security force. Another example could be from a regional foreign security force to a foreign security force with U.S. advisors. It could also be from a host-nation military force to a host-nation police force, with or without U.S. advisors.

Sustain

2-56. The sustain SFA activity occurs when the institutional capacity of the foreign security force has been developed to a point where it is self sustaining. It may continue to have SFA contact through combined exercises, educational opportunity exchange, intelligence sharing, and foreign military sales.

TYPES

2-57. The U.S. or other actors use three types of SFA—augmenting, partnering, and advising—to develop confident, competent, capable, committed, and credible FSF. These types of SFA define the relationship between outside actors and the foreign security force. They may be employed simultaneously, sequentially, or in combination. The progression and types of SFA are determined by the operational environment, the assessment of the FSF, and by resources available. Each of these types requires decidedly different requirements, objectives, and legal considerations.

Advising

2-58. Advising is the primary type of security force assistance. Advising is the use of influence to teach, coach, and advise (see chapter 6) while working by, with, and through FSF. This type of SFA relies on the ability of the advisor to provide relevant and timely advice to FSF. Advising helps FSF conduct independent decisionmaking and operations. Advisors may also provide FSF with direct access to joint and multinational capabilities such as air support, artillery, medical evacuation, and intelligence. However, the advisor and advisor team require proper manning and equipment to perform these secondary support functions while staying focused on advising. Multiple sources will pressure advisors as they perform their missions. For example, nearby maneuver units may have other priorities and may focus on accomplishing their missions with or without FSF.

2-59. Advisor teams require a clearly defined and structured chain of command under which to operate. This is not only for logistics and support, but also keeps the advisor focused on developing FSF. Advisor teams will often find themselves answering to their higher military assistance group, the brigade combat team they are attached to, and the host-nation unit with which they are embedded. A well-defined chain of command alleviates confusion concerning who tasks or monitors the team's progress as well as ensures constant and adequate team sustainment.

2-60. Advisors are not partners; U.S. forces act as partners. Advising and partnering are complementary but inherently different activities. Advising requires relationship building and candid discourse to influence development of a professional security force. Partnering incorporates training with combined operations to achieve the same SFA goals. Advisors perform partnership shaping functions, shape discussions with their counterparts, and create opportunities for the partner units.

PARTNERING

2-61. Partnering attaches units at various levels to leverage the strengths of both U.S. and foreign security forces. As a foreign security force's capabilities mature, the echelon and degree of partnering decrease. As the foreign security force conducts more autonomous operations, U.S. forces still provide quick reaction forces and other assistance as appropriate.

2-62. Partner units should establish combined cells for intelligence, operations, planning, and sustainment. While effective coordination is always required and initial efforts may require completely fused efforts, FSF should eventually build the capability and capacity to conduct all efforts autonomously. These combined cells have several functions. They support transparent operations and a comprehensive approach. They also enhance the relationships among U.S. and foreign security forces by demonstrating trust. Finally, they develop the capacity of FSF in key staff areas. However, combined cells are not without risk. Operations and plans always at risk of compromise and prudent precautions should be taken.

2-63. Another partnering technique establishes relationships among command and staff elements of U.S. and foreign security forces. It requires less reorganization. It also allows counterparts to understand each other since foreign personnel can observe U.S. personnel performing their duties. This construct works well when the foreign culture is extremely sensitive to maintaining honor. In this case, foreign personnel can observe without having to expose their lack of knowledge or appear subordinate.

2-64. Unit partnerships do not replace advisor roles or functions. If partnering and advising are used in combination, it forms a three-part relationship amongst FSF, advisors, and the partner units. Partner units should look to the advisor to identify, shape, and facilitate operational partnering opportunities and training events. Advisors support U.S., coalition, and partner unit objectives, but, depending on the operational phase, the partner unit may support advisors or advisors may support the partner unit. Therefore, some level of advisor skills training should be included in the partner unit training program if those units will be conducting SFA activities.

AUGMENTING

2-65. Augmenting is an arrangement where FSF provide individuals or elements to combine with U.S. units, or U.S. individuals or elements combine with FSF. Augmentation improves the interdependence and interoperability of U.S. and foreign security forces. Augmentation can occur at many levels and in many different forms. For example, a U.S. squad can be augmented with host-nation individuals, a U.S. company can be augmented with a host-nation platoon, or a U.S. battalion can be augmented with a company from a foreign security force. Similarly, augmentation can be of short duration for a specific operation or of a longer duration for an enduring mission. Augmenting immerses FSF in a U.S. environment to provide language and cultural awareness to the U.S. unit. U.S. forces can also augment FSF.

COMBINING TASKS, ACTIVITIES, AND ASSESSMENT

2-66. Execution of successful SFA links its activities, tasks, and assessment. Figure 2-2 depicts how to arrange SFA activities, tasks, and assessment in time and purpose. The tasks that occur in one SFA activity will normally continue throughout operations. For example, organize, train, equip, rebuild and build, and advise and assist tasks will continue throughout the SFA activities of employment, transition, and sustain.

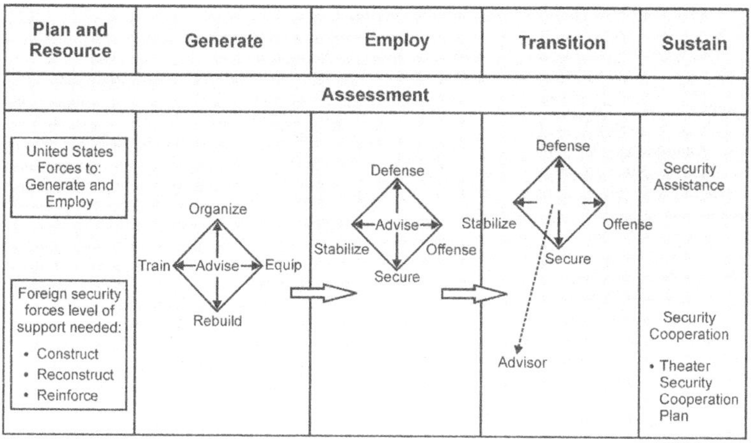

Figure 2-2. Security force assistance tasks, activities, and assessment

Plan and Resource

2-67. The plan and resource SFA activity determines the level of support and number of U.S. forces needed. As all efforts in SFA this activity is part of a comprehensive approach. Planners assess whether they will be constructing a foreign security force from the ground up, reconstructing a foreign security force based on existing capabilities and structure, or merely reinforcing an existing foreign security force. Understanding what capability and capacity the foreign security force should have and what capability and capacity they currently have is vital. Commanders and planners can then determine the size and structure, resources, and funding for the U.S. unit or units that will facilitate or assist the generation of FSF. Chapters 3 and 4 discuss planning and execution considerations for SFA.

Generate

2-68. In the generate SFA activity, U.S. forces focus on organizing, training, equipping, and advising FSF. During generation of FSF, the condition of the existing foreign security force's infrastructure and capabilities determine the assistance needed from U.S. forces. The generate activity adapts to the needs and requirements of the foreign security force. It can be designed to take the individual recruit from initial entry through to an employable unit, or it can augment an existing foreign unit with additional or enhanced capabilities. Like all SFA efforts, generate focuses on bringing a foreign security force's capabilities to a level where it can conduct its mission With help from advisors in the next phase, FSF can work toward independent and sustained employment. In large SFA efforts, an organization may be required to command and control this activity.

2-69. Once elements of FSF are trained, the focus shifts from organizing, training, equipping, and rebuilding to employing the foreign security force. However, not all FSF units and organizations are ready for employment at the same time. As one unit is employed, others may be part of the generation activity, while still others may be completing the transition activity. FSF may be employed for offensive, defensive, or stability operations while other FSF may be employed against internal or external threats. The primary SFA task for U.S. forces is to advise and assist to facilitate the foreign security force operations.

Transition

2-70. The transition activity focuses on removing the U.S. elements and advisors once FSF are ready to conduct independent operations. All involved actors work in a comprehensive approach to assess when this can be achieved and to coordinate the transition itself. As FSF gradually progress toward the transfer of authority, close relationships forged between the FSF and their partners prove essential to sustainable development and successful transition. Genuine relationships engender trust and confidence, enabling increased responsibility and a well-executed transition process. These relationships also foster a clear understanding of command responsibilities and authorities. Such an understanding ensures FSF approach transition prepared to assume the full weight of their future role in the security sector. Success in developing FSF often depends more on relationships and personalities at the unit level than any other factor.

Chapter 3

Brigade Operations Process

This chapter examines operational considerations for the modular brigade augmented for security force assistance (SFA). It uses battle command and the operations process to examine these considerations.

BATTLE COMMAND

3-1. *Battle command* is the art and science of understanding, visualizing, describing, directing, leading, and assessing forces to impose the commander's will on a hostile, thinking, and adaptive enemy. Battle command applies leadership to translate decisions into actions—by synchronizing forces and warfighting functions in time, space, and purpose—to accomplish missions (FM 3-0). Commanders are central to battle command for brigade combat team (BCT) operations just as they are for any other operation. However, there are key considerations for successful battle command of a BCT conducting SFA. Most importantly, BCT commanders must work with their foreign counterparts, which may require facilitating operations when the foreign security forces (FSF) commander has the lead for the operations process.

UNDERSTAND

3-2. Successful SFA requires that BCT commanders thoroughly understand the operational environment. Conditions of the operational environment affect what SFA commanders provide. Commanders must understand the conditions that make up the current situation, including the relevant population, other actors, and FSF. From this understanding, commanders visualize desired conditions that represent a desired end state. Commanders must understand the capacity and capabilities of foreign security. The host-nation or regional security organization helps to determine desired conditions associated with the foreign security forces. After envisioning a desired end state, commanders then conceptualize how to change the current situation to the desired end state.

3-3. BCT commanders must understand the host-nation's security and political concerns. Security concerns focus on internal and external threats, their capability to affect the operational environment, and the foreign security force's strength and capability to counter those threats. Politically, the considerations associated with the security sector are the legitimacy of the foreign security force, FSF capability and capacity, host-nation or partner organization policy, and the ability to support the host-nation and U.S. Government objectives.

3-4. Commanders use the information obtained from intelligence preparation of the battlefield and intelligence, surveillance, and reconnaissance (ISR) to determine objectives and requirements for assessing and mitigating risk. Risk applies to how well FSF, U.S. Government, host-nation, partner organization, and international organizations can tolerate changes in the operational environment, as well as the challenges and conditions inherent to the operation.

VISUALIZE

3-5. *Commander's visualization* is the mental process of developing situational understanding, determining a desired end state, and envisioning the broad sequence of events by which the force will achieve that end state (FM 3-0). Visualization begins with assignment of the mission, which provides the focus for developing the commander's visualization. However, visualization must be continuous as SFA is dynamic. During planning, commander's visualization provides the basis for developing plans and orders.

3-6. The nature and extent of assistance provided to FSF are determined by this desired end state. SFA is often a slow process that does not lend itself to quick solutions. To ensure long-term success, commanders and host-nation or regional security organizations clarify early what conditions they desire. These conditions include the desired end state for FSF, which often include being—

- **Flexible**. FSF can accomplish broad missions. Flexibility requires an effective command and control organization appropriate for the partner or host-nation. This can include both internal and external threats.
- **Proficient**. The level of proficiency of FSF should include—
  - The capability of working effectively within a comprehensive approach.
  - The capability to integrate their operations with those of multinational partners.
  - Nonmilitary security forces competent in maintaining civil order, enforcing laws, controlling borders, securing key infrastructure, and detaining criminal suspects.
  - Nonmilitary security forces thoroughly trained in modern police ethos and procedures and which understand the basics of investigation, evidence collection, and proper court and legal procedures.
- **Self-sustained**. FSF can manage their own equipment throughout its lifecycle and perform administrative support.
- **Well led**. Leaders at all levels possess sound professional standards and appropriate military values; they are selected and promoted based on competence and merit.
- **Professional**. FSF are—
  - Honest, impartial, and committed to protecting and serving the entire population; they operate under the rule of law and respect human rights.
  - Loyal to the partner organization or host-nation government and serving partner or national interests; they recognize their role as the servants and not masters.
- **Integrated into society**. In the case of a host nation, FSF represent the host nation's major social groups; they are not seen as elements of just one faction.

3-7. These conditions may vary in different countries, but well-trained FSF should—

- Provide reasonable levels of security from external threats without threatening regional security.
- Provide reasonable levels of internal security without infringing on the populace's civil liberties or posing a coup threat to the host-nation government.
- Be founded upon the rule of law, including international principles of the rule of law.
- Be sustainable after joint and multinational forces depart.

3-8. When addressing internal threats, foreign military and police forces often perform unconventional functions. The military may have to fill an internal security role usually reserved for the police, or police may have forces so heavily armed that they would normally be part of the military. In the near term, FSF should—

- Integrate military capabilities with those of local, regional, and national police.
- Maintain the flexibility to transition to more conventional roles of external and internal defense based on long-term requirements.

3-9. To achieve this end state and intermediate objectives, the host nation should develop an internal defense and development (IDAD) strategy (with multinational assistance if required). This strategy should have a section on IDAD that addresses all aspects of SFA.

3-10. U.S. doctrine divides force development into domains: doctrine, organization, training, materiel, leadership and education, personnel, and facilities (DOTMLPF), which may be useful to analyze the IDAD. DOTMLPF elements are tightly linked, simultaneously pursued, and difficult to prioritize. Commanders monitor progress in all domains. Often military forces involved in such programs are tempted to impose their own doctrine and judgment on the host nation. Foreign security force doctrine, like the remaining DOTMLPF domains, must be appropriate to host-nation capabilities and requirements. Paragraphs 3-59 through 3-70 use DOTMLPF as a tool to assess FSF.

DESCRIBE

3-11. After BCT commanders visualize how to conduct SFA, they describe it to their staffs and subordinates to facilitate shared understanding of the mission and commander's intent. Commanders ensure subordinates understand the visualization well enough to begin planning. Commanders express their initial visualization as—

- Initial commander's intent.
- Planning guidance, including an initial concept of operations.
- Commander's critical information requirements.
- Essential elements of friendly information that must be protected.

Commander's Intent

3-12. Like any operation, SFA requires a clear, concise statement of what the force must do and the conditions the force must meet to succeed with respect to the enemy, terrain, and desired end state. In such an operation, the end state should include a description of desired conditions with respect to FSF capability and capacity. The commander's intent links the mission and concept of operations. It describes the key tasks and end state that, along with the mission, form the basis for subordinates' initiative.

3-13. A sample end state could discuss—

- How civilians perceive FSF as credible and legitimate and representing a legitimate authority.
- How FSF have the capability to support stability and security, retain the initiative and freedom of movement, disrupt enemy activities, influence the population, operate with other friendly forces, and sustain themselves.

3-14. Like any operation, how the BCT supports the higher organization's mission is vital. Additionally, the BCT supports FSF. Some sample key tasks are:

- Support the organization of a division of foreign infantry and two battalions of foreign police.
- Train leaders of FSF to create a cadre that can sustain their own training programs.
- Oversee the equipping of FSF and provide new equipment training.
- Support the rebuilding or building of infrastructure required to develop FSF.
- Advise FSF to develop the required capabilities and competencies associated with shared objectives.
- Assist the partner's security forces in meeting their operational requirements.

Planning Guidance

3-15. Commanders develop planning guidance for the staff from the commander's visualization. Planning guidance may be as broad or detailed as circumstances require; however, it must convey to the staff the essence of the commander's visualization. Due to the complex and dynamic nature of FSF, SFA often requires detailed planning guidance. Commanders ensure the staff understands their visualization while still permitting the staff to explore different options. Some considerations are:

- Friendly courses of action (COAs) with differing SFA forms and differing methods on how to augment, partner, or advise.
- Command and control architecture, intelligence architecture, and communications architecture to support dispersed SFA.
- Enemy COAs as they seek to defeat friendly efforts with respect to SFA activities.
- Enemy adaptation given the development of the foreign security force.
- Guidance for security during the conduct of SFA, including the BCT securing its advisor teams. The BCT determines the staging, needs, and ad hoc arrangements of advisor teams.
- ISR guidance to include identifying the focus of effort as it relates to SFA.

Commander's Critical Information Requirements

3-16. Commander's critical information requirements (CCIRs) will be developed for SFA just as for any other operation. These requirements are what commanders need to make decisions or the information they think is important for understanding the operational environment. These CCIRs should be compatible with the FSF commander's information requirements. CCIRs include priority intelligence requirements and friendly force information requirements.

3-17. Priority intelligence requirements are vital to support decisionmaking. During SFA, priority intelligence requirements must be focused on the enemy and the operational environment, not on FSF. U.S. forces do not focus intelligence efforts, such as collection, on the foreign forces they assist.

3-18. Friendly force information requirements provide the information that commanders need to understand the conditions of their own forces, as well as the conditions of the foreign security forces. For example, a BCT commander may direct the staff to better define a foreign security force's critical equipment shortages or maintenance requirements. During execution, commanders may focus on when FSF meet certain objectives or conditions. This information allows the commander and FSF commander to make adjustments, such as changes to the task organization, support requirements, or even the type of SFA.

Essential Elements of Friendly Information

3-19. An *essential element of friendly information* is a critical aspect of a friendly operation that, if known by the enemy, would subsequently compromise, lead to failure, or limit success of the operation, and therefore should be protected from enemy detection (FM 3-0). Essential elements of friendly information should be determined through a comprehensive approach. Information sharing and trust are vital elements of SFA. U.S. commanders must be aware of the inherent risks in this context and employ prudent risk mitigation efforts. Thus, successful SFA requires a balance between sharing information and risk mitigation efforts. Although essential elements of friendly information are not CCIRs, they have the same priority.

Intelligence, Surveillance, and Reconnaissance

3-20. *Intelligence, surveillance, and reconnaissance* is an activity that synchronizes and integrates the planning and operation of sensors, assets, and processing, exploitation, and dissemination systems in direct support of current and future operations. This is an integrated intelligence and operations function. For Army forces, this activity is a combined arms operation that focuses on priority intelligence requirements while answering the commander's critical information requirements (FM 3-0). For the brigade, these efforts will often support FSF. Moreover, the BCT must have means that allow effective intelligence collaboration, such as an intelligence fusion cell with the participation of all appropriate actors.

3-21. ISR efforts focus on the enemy and operational environment within the context of SFA. ISR efforts in SFA are not used on the foreign security force. Rather, U.S. forces assess the foreign security force within a comprehensive approach.

DIRECT

3-22. BCT commanders direct all aspects of operations, although a comprehensive approach, including FSF and other appropriate actors is fundamental to successful SFA. The commander's direction takes different forms during planning, preparation, and execution. Commanders make decisions and direct actions based on their situational understanding, which they maintain by continuous assessment and coordination with FSF. SFA is directed by—

- Preparing and approving plans and orders with FSF.
- Assigning and adjusting missions, tasks, task organization, and control.
- Positioning units to maximize their support to FSF, anticipate threat actions, or create or preserve options.
- Positioning key leaders to ensure observation and supervision of FSF at critical times and places.

- Adjusting SFA and other support priorities and allocating resources based on FSF priorities, as well as opportunities and threats.
- Changing support arrangements, either internally or with respect to FSF.

ASSESS AND LEAD

3-23. Effective battle command requires commanders to continuously assess and lead. Assessment helps commanders to better understand current conditions and determine how the operation is progressing and how best to support FSF. Based on an assessment that is the product of a comprehensive approach, commanders modify plans and orders to better accomplish the mission. This may include changing the type or combination of SFA, support, sustainment, or task organization. If their assessment reveals a significant variance from the original shared commander's visualization, a comprehensive approach must be used to reframe the visualization and develop a new plan.

3-24. Commanders in a BCT conducting SFA lead by example and presence, as Army commanders do in any operation. However, commanders have the additional challenge of leading by, with, and through FSF. Command posts for BCTs conducting SFA have similar challenges in that they should be integrated headquarters as much as practical.

OPERATIONS PROCESS

3-25. As they would in any other mission, BCT commanders conducting SFA balance their time and the staff's time and resources among four major activities in a continuous learning and adaptive cycle called the operations process. The *operations process* consists of the major command and control activities performed during operations: planning, preparing, executing, and continuously assessing the operation. The commander drives the operations process (FM 3-0). While simple in concept, the operations process is complex in execution. The operations process activities may be sequential or simultaneous.

3-26. Figure 3-1 depicts the operations process activities as a cycle that is driven by battle command. This cyclic process applies equally to a BCT conducting major combat operational theme or operations within an irregular warfare operational theme. During the operations process involving SFA, commanders consider the presence and inclusion of other actors, including FSF. Policy, the operational environment, and the forms of SFA determine how U.S. forces, FSF, and other actors interact in the operations process.

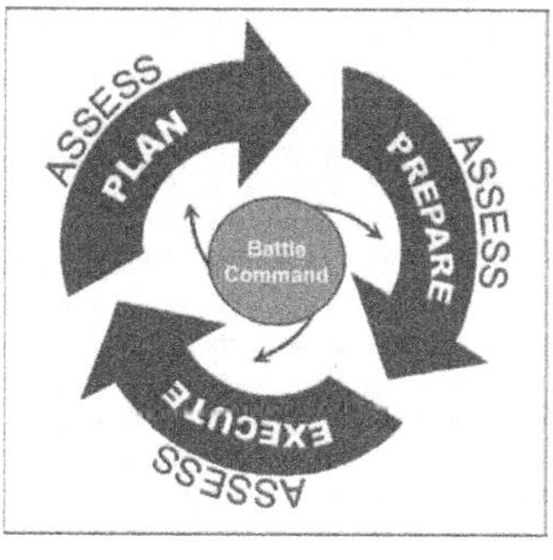

Figure 3-1. The operations process

PLAN

3-27. Planning for SFA, like any other operation, begins either with the anticipation of a new mission or the receipt of mission as part of the military decisionmaking process. *Planning* is the process by which commanders (and the staff, if available) translate the commander's visualization into a specific course of action for preparation and execution, focusing on the expected results (FM 3-0). SFA requires a comprehensive approach to ensure the U.S. commander's visualization synchronizes with policy and the foreign commander's visualization. SFA is the major requirement for a BCT conducting SFA, but the mission may require the brigade to perform other tasks.

3-28. Continuous and open to change, planning for SFA includes identifying how to best assist the FSF and developing a sequence of actions to change the situation. This planning must occur within a comprehensive approach that includes FSF, U.S. forces, and other actors. Similarly, this plan must be nested within policy, IDAD strategy, the campaign plan, and any other higher-echelon plans. It also involves anticipating consequences of actions and developing ways to mitigate them.

3-29. The plan, which includes the commander's intent, provides a common understanding to U.S. and foreign security forces on the actions to take. Plans and orders also provide decision points and branches that anticipate options that enable the force to adapt as the operation unfolds. This is especially important for SFA, as these operations tend to be prolonged efforts. BCTs conducting SFA often rotate before achieving all objectives. As a result, the planning team should establish objectives and milestones that can be achieved during the brigade's mission. These BCT objectives and milestones must support higher-echelon plans, including the campaign plan and IDAD strategy.

Functional Requirements

3-30. The composition of advisor teams is subject to objectives and conditions. For example, security conditions may require three armored vehicles, a crew-served weapon, and nine personnel to conduct independent movement. Another example involves the difference in the number of advisors. Newly generated FSF may require many advisors whereas experienced FSF may only require assistance in a few warfighting functions and consequently need fewer advisors. Understanding the functional requirements provides flexibility to the units assisting FSF. Three examples of determining functional requirements based on the understanding the operational environment include the BCT, a partnering or augmenting unit, and an advisor team.

Brigade Combat Team

3-31. A BCT headquarters has specific functions and requirements. These can include, but are not limited to—

- The functions of—
  - Carrying out SFA tasks and assisting FSF in support of objectives and the end state.
  - Facilitating partnerships between U.S. and foreign security forces.
  - Requesting, coordinating, and synchronizing resources.
  - Providing command and control of subordinate units and assigned advisor teams.
  - Coordinating with other commands and adjacent units.
  - Coordinating with other participants integrated within the comprehensive approach.
  - Developing, implementing, and synchronizing information engagement in assigned areas.
  - Developing relationships and an understanding with appropriate actors involved in SFA.

3-32. The requirements of—

- Having enough staff to handle the communications traffic generated by conditions.
- Being able to analyze information to aid commander and subordinate elements.
- Being able to articulate resource requirements in support of subordinate elements and efforts to conduct SFA tasks.
- Being organized and equipped to conduct operations and support and sustain FSF.
- Including regionally oriented host-nation officers to provide regional expertise.

Partnering or Augmenting Unit

3-33. Functions and requirements depend on the degree and level of partnering or augmenting. No prescriptive way to partner or augment exists; it is based on specific conditions and subject to the needs of policy, higher headquarters guidance, and needs of the FSF. A BCT may designate one battalion to partner or augment while designating other battalions to advise. A partnering or augmenting unit conducting SFA has specific functions and requirements. These can include, but are not limited to—

- The functions of—
  - Partnering with foreign security under the BCT level.
  - Increasing capability, capacity, competency, confidence, and commitment of FSF by conducting combined tactical operations.
  - Providing feedback on performance of FSF and certain shortfalls to embedded advisors and BCT headquarters.
  - Conducting sustainment and medical training with FSF at their home station or on operations.
  - Reporting conditions in the operational environment.
- The requirements of—
  - Having U.S. forces that are capable, competent, committed, and confident.
  - Being interoperable with FSF.
  - Being available for an appropriate period.

Military Advisor Team

3-34. The functions and requirements of a military advisor team are based on conditions—requirements, availability, and footprint. A military advisor team conducting SFA has specific functions and requirements. These can include, but are not limited to—

- The functions of—
  - Under the BCT headquarters, advising, training, and assisting FSF to which it is assigned.
  - Increasing capability, capacity, competency, confidence, and commitment of FSF by providing advice and support during battalion level and higher operations.
  - Assessing partner leaders, staff, and certain shortfalls to BCT headquarters.
  - Conducting sustainment training with FSF at their home or on operations.
  - Reporting on conditions in the operational environment.
- The requirements of—
  - Being manned with trained personnel who are properly equipped so as to be capable of executing their mission in current conditions.
  - Providing personnel with rank, education, and experience corresponding to the echelon and type of unit being advised.

Support and Sustainment Requirements

3-35. Support and sustainment requirements play a vital role in COA development (see chapter 6). The BCT may operate across a broad geographical area when conducting or supporting SFA. Partnering may be on certain levels with FSF and may support numerous embedded advisor teams or other actors. Since FSF often have varied relationships with higher echelons and institutions, the BCT may be involved vertically and spatially. These considerations may require more support and sustainment requirements than an initial analysis might indicate. Support and sustainment considerations include, but are not limited to—

- Distribution of FSF to which U.S. forces are providing SFA.
- The capability of FSF to provide support, sustainment, and medical functions.
- The technical capability and interoperability of U.S. and foreign security forces.
- The infrastructure that supports SFA.
- Overall security conditions where SFA occurs.

- The available contracting support.
- The ability of the environment to accept and absorb the U.S. forces' footprint.

Determining a Course of Action

3-36. COAs for SFA require an understanding of the breadth and width of the security sector. They may have to account for the roles ranging from the individual soldier or policeman to the ministerial level. COA development examines the roles other actors, for example, a gendarme or paramilitary force, in the security sector. Because the United States does not have a gendarme, U.S. forces may not fully appreciate its importance.

3-37. COAs for SFA also need to be flexible and inclusive. Commanders consider the following when examining a partner's security sector:

- The status and challenges of the host-nation security sector or regional security organization.
- The composition and disposition of the partner's security sector.
- Supporting institutions.
- Links among ministries, institutions, and components of the security sector.
- Partner capability and capacity shortfalls and impediments to their development.
- Environmental requirements to enable the building of security capability and capacity.
- Functional requirements to assist building capability and capacity.
- The impact of other actors within the comprehensive approach.

3-38. Once planners analyze the security sector, they can develop COAs that best combine SFA tasks, activities, and types (see chapter 2). While the advising type of SFA can be of any duration, partnering and augmenting tend to be enduring if used.

3-39. All SFA types tend to be more effective if they are long-term efforts. Advising fosters a more personal relationship and infers a stronger commitment to see the partner succeed. Partnering and augmenting include U.S. and foreign security forces conducting operations together. There are levels of partnering. Different advisory relationships use advisor teams embedded within various partner echelons and units, use one advisor team to advise multiple entities, or have a senior leader visit a partner unit to give advice.

3-40. Using any form or combination of forms depends on conditions (such as the level of competency), capability of FSF, and resources available. Policy, guidance, or formal agreements can also help commanders determine which form to use. Historically, a combination of forms appears to be the most effective. However, combining forms often proves more resource intensive and affected by time, geography, and the number of forces. COAs should remain flexible with respect to the form of SFA. Some example changes in conditions are the increased capability of one element in the security sector over another, the desire to increase one element over another based on partner objectives, or the increased targeting of one element over another by the enemy.

3-41. If a desired COA is not feasible given the scope of the security sector or due to other considerations, it may be useful to sequence SFA forms with portions of FSF. For example, a COA that affects smaller portions of the FSF, but in the right area and at the right time, may do more for achieving an objective or end state than a COA that attempts to support all FSF at one time.

3-42. BCTs are task-organized and augmented to accomplish a specific mission. How BCTs are specifically task-organized and augmented is based on the phase and types of SFA that policy and higher headquarters have directed. Chapter 4 describes augmentation in detail.

3-43. Attached units or units under operational control are task-organized under different headquarters as necessary to accomplish assigned and implied SFA missions. Deciding to task-organize specific formations for SFA may reflect previously approved contingency plans or result from current planning efforts. As these missions end, units return to their parent headquarters or are further task-organized by the controlling headquarters.

3-44. As with other operations, gaining commanders task-organize for SFA by designating operational control, attached, or support relationships to another unit or headquarters. Within each BCT, the brigade commander task-organizes all organic, assigned, attached, or operational control units to best suit the operational environment and types of SFA. Commanders may choose to use a support relationship (direct support, reinforcing, or general support reinforcing) among units of the BCT instead of a command relationship. This can also apply to relationships with foreign units.

3-45. SFA is often highly dispersed, as BCTs may have individuals or units assisting FSF over a wide area. This dispersion increases communications and sustainment requirements for the BCT. The nature of SFA also requires BCTs to help FSF meet requests for assets. To do so, the BCT and other units may need to task-organize to provide tactical support or sustainment. Dispersion and the nature of SFA often necessitate a decentralized leadership approach to facilitate effective and rapid support of FSF.

3-46. BCT commanders use the three types of SFA—advising, partnering, or augmenting—to accomplish the mission. The exact nature of a unit's assigned mission, the operational variables, and the mission variables drive modifications to headquarters and maneuver units. For example, a commander may need a unit tasked with advising, a unit tasked with partnering, and a unit tasked with augmenting. In this case, the unit tasked to advise would provide teams for the appropriate FSF while other foreign units continue their traditional missions. The unit that is partnering may provide the host-nation staff to assist the FSF headquarters. The third unit may provide squads to augment foreign platoons. Units may request additional support, sustainment, and medical support. However, BCTs conducting SFA must retain a reserve to conduct full spectrum operations.

3-47. Based on the overall mission analysis, including an analysis of functions and requirements, commanders and staffs begin to determine how to task organize. Due to the level of detail that may be involved, task organization should begin early in the military decisionmaking process, especially during COA development. No standard task organization exists since conditions vary. For example, a squad of infantry and three additional leaders may work as an advisor team for one mission, where another mission may require an advisor team with specialists for different warfighting functions. The art in SFA involves determining where, when, and if to allocate resources; it involves either breaking up or simply moving personnel and units from one foreign unit to another.

3-48. The techniques used by BCT commanders and staffs to allocate other resources and support are just as valid in SFA as in other types of operations. For example, if it is determined that the most important effort is advising, then the BCT commander may weight that effort with additional resources, such as communications equipment or experienced personnel.

3-49. BCTs emphasize unity of effort with FSF. Ideally, interoperability and shared resources extend to the lowest level. This means having one tactical operations center, dining facility, clinic, motor pool, and so on. Not only does this use resources more efficiently, but it also facilitates understanding, trust, and rapport between U.S. and foreign security forces. For Soldiers to effectively conduct SFA work by, with, and through FSF requires a close rapport and high level of trust.

3-50. Part of COA development for SFA includes an analysis of where to focus the brigade's subordinate units. This analysis includes determining which subordinate units support key elements of FSF. The brigade's subordinate units may support provincial reconstruction teams or transition teams that are not organic to the brigade. Another method is for the brigade may provide organic forces to make up transition teams or the military portion of provincial reconstruction teams or transition teams. Transition teams may include, but are not limited to police, border, point of entry, or military transition teams. Figure 3-2 depicts an example for a brigade providing forces as part of a provincial reconstruction team as well as providing the other transition teams. Figure 3-2 is an example of how a brigade may focus its subordinate units. Additionally, the teams are only examples. All of these teams may not be necessary for all brigades, and some brigades may require other types of advisor teams.

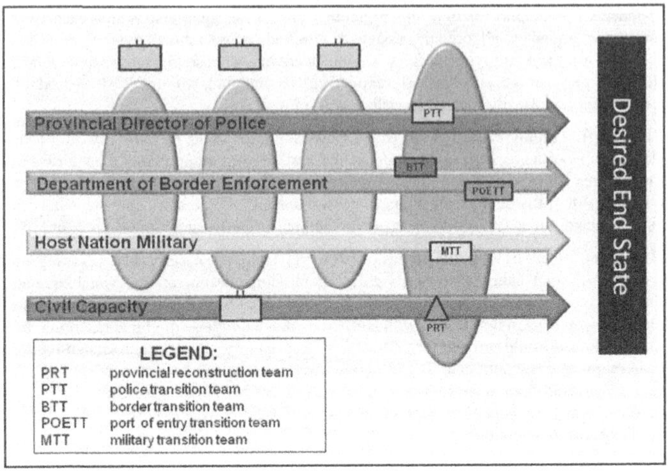

Figure 3-2. Example of brigade focus for subordinate units

PREPARE

3-51. *Preparation* consists of activities performed by units to improve their ability to execute an operation. Preparation includes, but is not limited to, plan refinement; rehearsals; intelligence, surveillance, and reconnaissance; coordination; inspections; and movement (FM 3-0). Like any operation, preparation for SFA creates conditions that improve friendly forces' opportunities for success. It facilitates and sustains transitions, including those to branches and sequels, which are of vital importance for the often dynamic operational environment for SFA. Chapter 5 covers preparation for units conducting SFA, including the BCT.

EXECUTE

3-52. *Execution* is putting a plan into action by applying combat power to accomplish the mission and using situational understanding to assess progress and make execution and adjustment decisions (FM 3-0). Execution of SFA focuses on supporting the needs of the FSF and, like many other operations, is often characterized by phases and transitions between phases. Chapter 4 examines BCT phases for SFA.

Acceptance of Prudent Risk

3-53. SFA, like any other operation, requires prudent, disciplined risk-taking focused on winning rather than preventing defeat. Because uncertainty exists in all operations, every decision involves risk. Among key elements of the art of command are deciding how much risk to accept and minimizing the effects of accepted risk (see FM 6-0). Due to SFA's inherent collaborative nature, determining prudent risk is complex and requires a comprehensive approach.

3-54. Risk reduction measures identified in planning add to the plan's flexibility during execution. A flexible plan can mitigate risk by partially compensating for a lack of intelligence. Again, SFA requires a thorough comprehensive approach to analyzing and agreeing upon risk reduction measures.

Execution Keys to Success

3-55. History and recent operations point to several keys to success for BCTs executing SFA:

- Communication is vital, especially when the BCT shares its operational area with other entities that have cultural differences and lack of communication. It can include FSF or multinational forces, as well as U.S. forces with another mission. For example, an advisor team can develop effective relationships with its foreign security force and local authorities only to have another unit conduct an operation that unhinges those relationships.
- BCTs establish close and continuing relationships with all advisor teams, other actors operating in their operational area, and foreign area officers with local or regional expertise. In addition to advisor teams assigned to military units in direct contact with the BCT, advisors may embed with local police, border, and other forces.
- BCTs establish close and continuing relationships with all foreign units (military, police, and others) operating in their operational areas.
- BCTs establish close and continuing relationships with all political entities and actors in their operational areas.
- Before each operation, as part of the military decisionmaking process, the BCT considers second- and third-order effects of the operation. The team coordinates the commander's intent with advisor teams of FSF and other units in the operational area.
- Unit leaders develop cultural awareness and use this awareness so that operations and relationships achieve the desired end state. In short, the closer a maneuver commander works with advisor teams and the more they interact with local political and cultural leaders, the better the overall chances of mission success.

ASSESS

3-56. The operations process and effective battle command require commanders to continuously assess and lead. Assessment helps commanders to better understand current situations and how SFA is progressing. *Assessment* is the continuous monitoring and evaluation of the current situation, particularly the enemy, and progress of an operation (FM 3-0). Assessment is conducted during planning, preparation, and execution and is evaluated against measures of effectiveness and performances to make decisions and adjustments (see FM 6-0).

3-57. Assessment of SFA requires a comprehensive approach that analyzes the overall operational environment and focuses on FSF and their impact. SFA assessment is performed in close collaboration with the U.S. country team, the host-nation government, and other multinational partners. These partners continually assess the security situation and its influence on other operations. From the assessment, planners develop short-, mid-, and long-range plans. As circumstances change, so do the plans. An insurgency might require employing host-nation forces at earlier stages of development. Some existing security forces may be so dysfunctional or corrupt that the organizations ideally should be disbanded rather than rehabilitated. In some cases, commanders will need to be replaced before their units will become functional. Political realities, however, may make both prospects undesirable. Minimizing and marginalizing particularly corrupt units or leaders while focusing on other units can be effective in this case.

3-58. While every situation differs, BCT leaders assess the following factors throughout planning, preparation, and execution of the operation:

- The operational environment.
- The civil considerations of the operational environment (see appendix B of FM 6-0).
- The core grievances and the threat (possibly including the prerequisites of insurgency).
- FSF elements.
- Methods, successes, and failures of FSF security efforts.
- State of training at all levels, and the specialties and education of leaders.

- Equipment and priority placed on maintenance.
- Logistic and support structure and their ability to meet the force's requirements.
- Extent of acceptance of ethnic and religious minorities.
- Laws and regulations governing the security forces and their relationships to national leaders.

Initial Assessment

3-59. An initial comprehensive assessment offers the BCT a baseline for establishing the scope of effort required. This assessment includes a troop-to-task analysis that determines the type and size of forces needed. FSF may require complete reestablishment or only assistance to increase capacity. They may completely lack a capability (such as internal affairs, corrections, or formal schools for leaders) or they may require temporary reinforcement. As with other military operations, efforts to assist FSF should reinforce success, not failure. For example, instead of building new police stations in every town, improve the good stations and use them as models for other organizations.

3-60. BCT leaders need to make decisions on what shortfalls to address first. The extent of the threats combined with resource limitations inevitably forces commanders to set priorities. However, making efforts to legitimize the host-nation leaders by encouraging them to make the decisions often proves more successful.

3-61. A key effort for SFA is the initial assessment of the DOTMLPF and policy capabilities of a foreign security force. During the early phases of a campaign, a geographic combatant command receives an initial assessment or a unit operating in the area has an assessment of FSF. The incoming BCT updates this assessment as it begins SFA.

3-62. Assessments require objective and subjective reporting, which may require a classification level. At a minimum, a BCT should assess—

- Doctrine.
- Organization.
- Training.
- Materiel and equipment.
- Leadership.
- Personnel.
- Command, control, communications, and intelligence.
- Operational effectiveness.

3-63. The doctrine assessment provides the BCT with an analysis on how the FSF operate. This assessment could focus on formal doctrine or more localized practices.

3-64. The organization portion of the overall assessment provides a perspective on how the FSF are organized. This includes the array of foreign institutions and units.

3-65. The level of training for FSF is a vital consideration. While this assessment is partly subjective, it should objectively examine individual and collective training strategies and mission-essential task lists. In the absence of mission-essential task lists, the BCT should compare FSF to U.S. equivalents.

3-66. The materiel and equipment assessment includes, but is not limited to, weapons, vehicles, equipment, and communications. It should be divided into equipment on hand, equipment readiness, and sustainment. Equipment on hand is quantitative and based on the type of equipment available to security operations. Equipment readiness reflects the operational readiness condition or serviceability of on-hand equipment. Sustainment readiness provides a qualitative and quantitative assessment that reflects how well FSF perform supply, maintenance, transportation, and medical functions. This includes facilities, systems, and structures.

3-67. The leadership assessment analyzes FSF commanders, officers, noncommissioned officers, and other leaders. This includes an assessment of staff members.

3-68. The foreign personnel assessment is quantitative and examines manning, staff manning, and administrative capabilities. It also examines officers, noncommissioned officers, and other personnel.

3-69. The command, control, communications, and intelligence assessment is qualitative and should reflect how well FSF conduct these functions.

3-70. The operational effectiveness assessment examines the foreign security force's ability to execute operations. This portion of the assessment should not be confused with the examination of training, as relatively high levels of training do not necessarily mean that FSF will perform in operations. This assessment is highly subjective.

Assessment During Operations

3-71. Assessment for the BCT conducting SFA should be part of a comprehensive approach. Assessment is directly tied to the commander's decisions throughout planning, preparation, and execution. Critical to effective assessment is developing criteria to evaluate progress toward task accomplishment, achievement of objectives, and attainment of the end state conditions. Assessment criteria are expressed as measures of effectiveness or measures of performance.

3-72. During operations, BCT and FSF commanders and staffs periodically assess the underlying framework of the plan itself. This involves reexamining the logic and assumptions used to develop the original plan, and it requires a deliberately conceived effort to learn as a part of the operation. BCT and FSF collaboration among higher, lower, and adjacent commanders, staffs, and other actors, backed up by qualitative and quantitative assessments, contributes to this learning. Based on this reexamination, commanders update their visualization, direct changes to the order, or develop an entirely new plan to adapt the force to better accomplish the mission.

Chapter 4

The Modular Brigade Augmented for Security Force Assistance

This chapter examines the typical phases for the brigade combat team conducting security force assistance, including a discussion of baseline augmentation by phase. The final section examines some of the subordinate units and elements of the modular brigade augmented for security force assistance.

THE BRIGADE COMBAT TEAM

4-1. The brigade combat team (BCT) is the cornerstone for Army modularity and is designed to operate at the tactical level across the spectrum of conflict. The modular brigade can be augmented, based on the requirements of the operational environment, with enabling assets and capabilities to support distributed security force assistance (SFA). However, the BCT augmented for SFA retains the capability to conduct full spectrum operations—offense, defense, and stability. Any of the three modular BCTs—heavy, infantry, or Stryker—can support SFA. Battalion-sized maneuver, fires, reconnaissance, and sustainment subordinate elements of the BCT support foreign security forces (FSF).

4-2. The gaining theater army force tailors all modular Army forces for use in major combat operations and contingencies. The theater army commander, working for the geographic combatant commander, determines the mix of forces and capabilities required for the BCT. This force tailoring can be done as part of the joint deliberate planning process or as a result of crisis action planning. Force tailoring includes providing BCTs with additional forces, personnel, or capabilities, including the possibility of an embedded provincial reconstruction team. Additional assets and capabilities can include, but are not limited to, command and control, communications, sustainment, engineering, military police, and intelligence.

4-3. In recent operations, the BCT has proven its versatility and capability to learn at the individual and collective levels. It supported other U.S. units conducting SFA and directly increased the effectiveness of FSF. The BCT has operated under varying levels of enemy activity, host-nation capabilities, objectives, and other conditions.

4-4. Because the BCT can operate in nonpermissive and permissive environments, it can conduct SFA across the spectrum of conflict. Thus, it can support civilian and military joint and multinational actors. The support can range from movement security and sustainment to augmentation of reconstruction teams and security for elections.

4-5. BCTs conducting SFA may support FSF development, assist FSF operations, and support and assist the development of host-nation institutions and infrastructure. While providing this support to the host nation, the BCT remains capable of conducting full spectrum operations independently.

BRIGADE COMBAT TEAM PHASES

4-6. If a force lacks the means to accomplish its mission in a single effort, then commanders normally phase the operation. A *phase* is a planning and execution tool used to divide an operation in duration or activity. A change in phase usually involves a change of mission, task organization, or rules of engagement. Phasing helps in planning and controlling and may be indicated by time, distance, terrain, or an event (FM 3-0). Phasing SFA assists in planning and controlling these dispersed and complicated operations. These phases often involve key changes in supporting SFA tasks and types employed.

4-7. BCT phases for SFA are initial response, transformation, and fostering sustainability. These phases mirror the stability operations framework described in FM 3-07 and are based on the operational environment. SFA can start in any phase or may even move to a previous phase due to changes in the conditions of the operational environment.

4-8. Figure 4-1 illustrates an SFA operation that includes all three SFA phases. The horizontal axis depicts an improving security situation, and the vertical axis depicts security capability. The example depicts an ideal situation, although this may not be the case in actual execution. Issues that arise require proper assessment and potentially the use of branches or sequels.

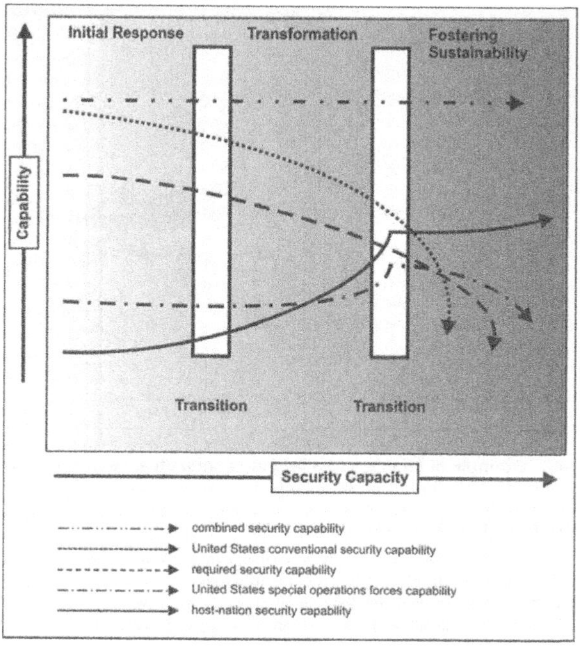

Figure 4-1. Example of phases of security force assistance

4-9. In figure 4-1, the combined security capability represents the overall security capacity of all security forces involved; it is the sum of U.S. conventional security, host-nation security, and the U.S. special operations forces security capabilities. This example assumes that the combined security capability exceeds the required security capability to effectively maintain a secure environment and counter any threat. The U.S. conventional security capability refers to tactical U.S. elements, such as the BCT. It does not refer to higher tactical or operational headquarters or support, sustainment, or medical assets. The U.S. special operations forces capability is depicted as declining in the end of the sustainment phase; as the FSF capability improves, the need for special operations forces declines. The host-nation or regional security organization and U.S. policy determine the end state for special operations.

4-10. For figure 4-1, a major change to enemy resources may require more FSF or additional. Drastic changes to the situation may even require a new operation plan. The figure focuses on capability, not forces, so host-nation security capability may increase without increasing the number or size of host-nation forces.

4-11. Figure 4-2 illustrates how the area of operations for a modular brigade augmented for SFA may change throughout the SFA phases. The differences between the initial response phase and the transformation phase may not change on the surface, but the BCT's relationship with FSF changes drastically. For example, FSF in the north met required conditions so that one of the U.S. brigades conducting SFA was no longer required. The responsibility of the other U.S. brigade thus expanded from providing assistance to one division to two divisions. This brigade's span of control and area of operations similarly expanded. The latter stage of the transformation phase can differ greatly. Areas of operation generally increase in size as they provide SFA to more FSF.

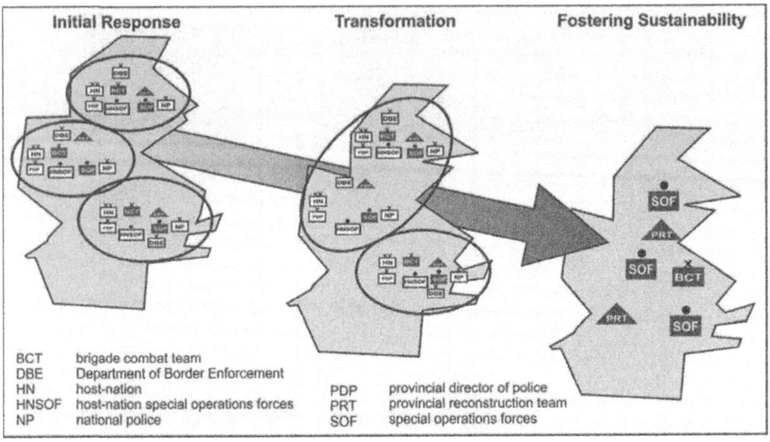

Figure 4-2. Example of changes to a brigade combat team's area of operations

AUGMENTATION BY PHASE SUMMARY

4-12. BCT augmentation is based on the operational environment and mission. As the three phases are based on the operational environment, they provide a baseline for augmentation. Potential augmentation may require military police, legal, public affairs, civil affairs, psychological operations, engineering, sociocultural experts, sustainment, and military transition team personnel. Table 4-1 illustrates a baseline for augmentation by phase for BCTs.

4-13. An embedded provincial reconstruction team is a key element of a BCT conducting security for assistance. The BCT has the lead on the following tasks: establish civil security, establish civil control (when approved by Congress), and develop and enable foreign security forces. An embedded provincial reconstruction team has the lead for the following tasks: support to economic and infrastructure development, restore essential services, and support to governance. Thus, the BCT and an embedded provincial reconstruction team are able to effectively support FSF and execute all five stability tasks.

Table 4-1. Baseline augmentation for modular brigade

CAPABILITY	INITIAL PHASE (1 HN DIV)	TRANSFORMATION PHASE (2 HN DIV)	TRANSFORMATION PHASE (3 HN DIV)
Police Provincial District	171 - MP Company 2 - MP Team (LTC/SGM) 3 - MP MAJ	171 - MP Company 4 – 2 x MP Team (LTC/SGM) 6 - MP MAJ	171 - MP Company 6 – 3 x MP Team (LTC/SGM) 9 - MP MAJ
Legal HN BDE HN DIV	2 - JAG Team (MAJ/SFC) 2 - JAG Team (CPT/SSG)	4 – 2 x JAG Team (MAJ/SFC) 4 – 2 x JAG Team (CPT/SSG)	6 – 3 x JAG Team (MAJ/SFC) 6 – 3 x JAG Team (CPT/SSG)
Public Affairs	2 - PA (Officer/NCO)	2 - PA (Officer/NCO)	2 - PA (Officer/NCO)
Civil Affairs & Psychological Operations	32 - CA Company = 5 CA Teams 70 – Tactical PSYOP Company = 9 TPTs	32 - CA Company = 5 CA Teams 70 – Tactical PSYOP Company = 9 TPTs	197 - CA Battalion (4 CA Companies) = 20 CA Teams 70 – Tactical PSYOP Company = 9 TPTs
Engineer	8 - FEST-A Team 2 - EOD (CPT/SFC)	8 - FEST-A Team 2 - EOD (CPT/SFC)	8 - FEST-A Team 2 - EOD (CPT/SFC)
Sociocultural Experts	8	8	8
Military Transition Team	56- MTT Augment	4 - MTT Augment	8 - MTT Augment
TOTAL	358	315	493
EPRT	~15-20	~15-20	~15-20
Sustainment Truck Squad Supply Platoon CCT	 21 personnel 32 personnel 4 personnel	 4 personnel	 4 personnel
FMST	4 personnel	4 personnel	4 personnel

CA	civil affairs	LTC	lieutenant colonel
CPT	captain	MAJ	major
CCT	contingency contracting team	MP	military police
DIV	division	MTT	military transition team
JAG	Judge Advocate General	NCO	noncommissioned officer
EOD	explosive ordnance disposal	PA	public affairs
EPRT	embedded provincial reconstruction team	SFC	sergeant first class
FEST-A	forward engineering support team – augmenting	PSYOP	psychological operations team
FMST	financial management support team	SGM	sergeant major
HN	host-nation	SSG	staff sergeant
HTT	human terrain team	TPT	tactical psychological operations team

INITIAL RESPONSE PHASE

4-14. The initial response phase seeks to assist FSF stabilize the operational environment in a crisis state. This phase occurs during or immediately after a conflict where the operational environment prevents civilian personnel from operating effectively. The operational environment is typified as nonpermissive.

Thus, the objective of this phase is to improve the security situation, reducing the threat to the populace and creating the conditions that allow civilian personnel to safely operate.

4-15. SFA in the initial response phase is normally required when FSF lack the capability or capacity to provide the required level of security. This phase often requires SFA efforts to help generate and train or assist new and existing FSF. This phase may require a combination of the types of SFA and considerable support, sustainment, and medical resources. BCT activities during the initial response may have to be conducted with multinational major combat operations, to include providing a safe, secure environment for the local populace.

4-16. The BCT's SFA efforts during this phase focus on improving the foreign security force's capability and capacity so all security forces—U.S., other, and foreign security forces—provide a secure environment and reduce the threat. Figure 4-3 illustrates the BCT's span of control. The figure also depicts three BCTs' operational areas in the initial response phase. As security conditions improve, transition to the transformation phase begins.

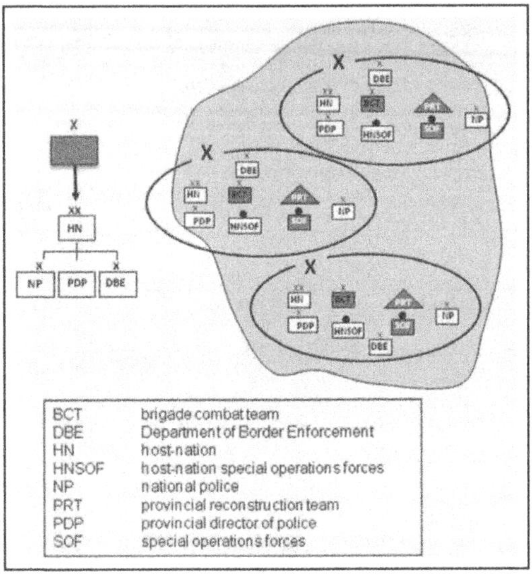

Figure 4-3. Example of initial response span of control and area of operations

4-17. The BCT requires augmentation for the initial response phase. These embedded units provide key enabling capabilities for the brigade, supported FSF, and other supported actors. The embedded provincial reconstruction team is one of the most important of the brigade's augmentation, as it provides interim civilian expertise in developing local institutions to take the lead in national governance, the provision of basic services, fostering economic development, and enforcement of the rule of law. Column two of table 4-1 summarizes augmentation during the initial phase. Although each brigade's mission and conditions in its area of operations are unique, the baseline augmentation for the modular brigade operating in this phase is as follows:

- Embedded provincial reconstruction team.
- Forward engineering support team-augmenting.
- Explosive ordnance disposal team.
- Sociocultural experts.

- Military police company (provincial police transition team).
- Military police team (district police transition team).
- Three military police majors (district police transition team).
- Two judge advocate general teams.
- Civil affairs company.
- Tactical psychological operations company.
- Public affairs team.
- Contingency contracting team.
- Financial management support team.
- Military transition team (specific augmentation personnel include two lieutenant colonels, thirty-one majors, and twenty-three master sergeants).

TRANSFORMATION PHASE

4-18. In the transformation phase, the BCT seeks to assist FSF to stabilize the operational environment in a crisis or vulnerable state. The operational environment in this phase is more permissive than the initial response phase; however, military forces will often be required to provide security to some actors. Activities in this phase normally include a broad range of post-conflict reconstruction, stabilization, and capacity-building efforts, which the embedded provincial reconstruction team is essential for long-term success. Objectives in this phase include continuing efforts to improve the security situation, reducing the threat to the populace, building host-nation capacity across the stability sectors, and facilitating the comprehensive approach to assist FSF.

4-19. The transformation phase represents a broad range of SFA activities that the BCT could perform to support FSF. The initial response phase differs from the transformation phase in the foreign security force's capability to provide for a safe and secure environment. More specifically, FSF have a level of proficiency so they no longer need a permanent U.S. and foreign security forces relationship for tactical operations. However, they may still need full-time advisors and support, sustainment, and medical assistance. The embedded provincial reconstruction team will continue to play a vital role in assisting governance and development efforts throughout this phase.

4-20. Figure 4-4 illustrates the BCT conducting SFA in the transformation phase. In this example, the BCT's span of control expands from assisting one foreign division to assisting two foreign divisions. The figure also depicts how a BCT's operational area has similarly expanded. SFA activities in this phase seek to establish conditions so the host nation's security sector can provide a secure environment with its own security forces.

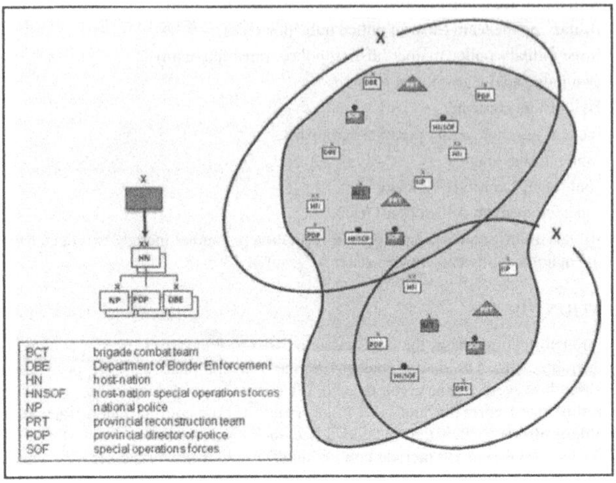

Figure 4-4. Example of transformation phase span of control and area of operations

4-21. The BCT requires augmentation for the transformation phase, like the initial response phase. It often requires similar augmentation due to its greater span of control and area or operations. Column three of table 4-1 summarizes augmentation during the transformation phase with two foreign divisions. The baseline augmentation for the brigade in this phase is as follows:

- Embedded provincial reconstruction team.
- Forward engineering support team-augmenting.
- Explosive ordnance disposal team.
- Sociocultural experts.
- Military police company (provincial police transition team).
- Two military police teams (district police transition team).
- Six military police majors (district police transition team).
- Four judge advocate general teams.
- Civil affairs company.
- Tactical psychological operations company.
- Public affairs team.
- Contingency contracting team.
- Financial management support team.
- Military transition team (specific augmentation personnel include two lieutenant colonels and two master sergeants).

4-22. As conditions of the operational environment improve, the BCT may require a transition to assist three foreign divisions, rather than two divisions as depicted in figure 4-4. This represents a significant expansion of the span of control and area of operations, with a corresponding expansion of required augmentation. Figure 4-5 depicts the BCT supporting three foreign divisions.

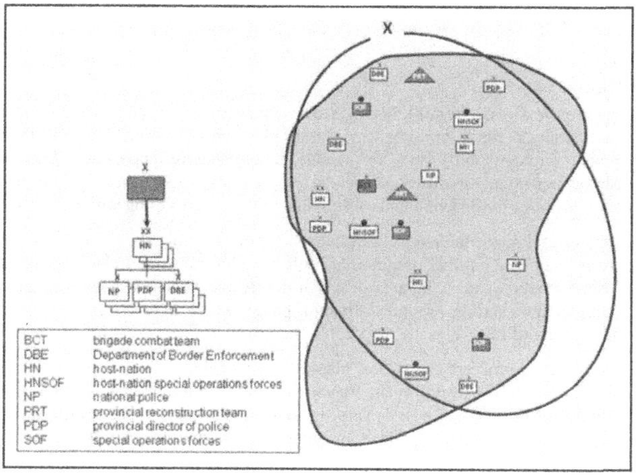

Figure 4-5. Example of a three-division span of control and area of operations

4-23. As in other phases, the BCT requires augmentation to support three divisions. Column four of table 4-1 summarizes augmentation during the transformation phase with three foreign divisions. The baseline augmentation for the brigade in this phase is as follows:

- Embedded provincial reconstruction team.
- Forward engineering support team-augmenting.
- Explosive ordnance disposal team.
- Sociocultural experts.
- Military police company (provincial police transition team).
- Three military police teams (district police transition team).
- Nine military police majors (district police transition team).
- Six judge advocate general teams.
- Civil affairs company.
- Tactical psychological operations company.
- Public affairs team.
- Contingency contracting team.
- Financial management support team.
- Military transition team (specific augmentation personnel include four lieutenant colonels and four master sergeants).

Fostering Sustainability Phase

4-24. In this phase, the focus of SFA continues to shift toward assisting institutions required to sustain FSF. This phase encompasses long-term efforts to assist FSF. In this phase, FSF are capable and viable. They conduct independent operations and can provide a safe, secure internal environment. While the brigade may be initially required during this phase, it is normally withdrawn during this phase. The determination to withdraw the brigade is based on the policy and conditions of the operational environment. Special operations forces, provincial reconstruction teams, and other forces may remain to support a theater security cooperation plan. Thus, remaining SFA missions are normally special operations forces missions, not a BCT mission (special operations forces will normally be involved in all phases of SFA).

Transitions

4-25. SFA can require a vast array of transitions. These transitions are conducted according to the tactical, operational, and strategic conditions identified during planning (which must support the internal defense and development strategy and using an overall comprehensive approach). Transitions can occur simultaneously or sequentially in different levels or war and in separate echelons. This includes potentially having tactical transitions for different units within one brigade's area of operations.

4-26. There are several major transitions. First, the supported unit may be the U.S. BCT in the beginning of an initial response phase but the foreign security forces will transition to be the supported unit. At this point in the transformation phase, the area in which the BCT conducts SFA will expand. This expansion can occur multiple times during the transformation phase, which is based on conditions, especially the capability and capacity of FSF.

4-27. Operations never proceed exactly as planned. To facilitate flexibility, commanders incorporate branches and sequels into the overall plan. Visualizing and planning branches and sequels are important because they involve transitions. Unless planned, prepared for, and executed efficiently, transitions can reduce the tempo of the operation, slow its momentum, and surrender the initiative to the adversary.

SUBORDINATE UNITS

4-28. The modular brigade augmented for SFA has subordinate units whose sole focus is working with FSF. These advisor teams may be formed from brigade organic resources, external augmentation, or a combination. These teams optimally are embedded with the counterpart unit, or they may reside on a U.S. camp and commute to FSF they support. The method depends on U.S. policy, direction from higher headquarters, the conditions of the operational environment, and capacity of the foreign security force's camps to accommodate the U.S. forces.

4-29. The military transition team (MTT) is a key subordinate unit for the brigade conducting SFA. The MTT mission is to assist FSF units. Figure 4-6 illustrates the relationship between a brigade conducting SFA and a MTT assisting a foreign brigade. In this example, the brigade has a battalion and a division MTT as subordinate units, although it would have other subordinate units. The brigade MTT is subordinate to the battalion and has operational control of one of the parent U.S. battalion's company teams.

4-30. When the brigade provides organic forces to form the basis for the brigade MTT, the company team is the foundation of a brigade MTT. The company team will be augmented with additional personnel and assets. The U.S. brigade must facilitate the MTT operations, such as force protection, transportation, communications, and reachback.

4-31. As companies do not have organic intelligence analysts, personnel must be used to form company intelligence support teams (COISTs). A COIST, an ad hoc organization, assists the MTT and its FSF. Ideally, the COIST should consist of six individuals, which will allow for 24 hour operations and the depth in the team to both integrate with the staff and to successfully support the company's mission. The team should consist of an officer, a noncommissioned officer, and four Soldiers, although the cell can be minimally manned with four personnel. The company intelligence cell works most effectively if its work complements the battalion S-2's work.

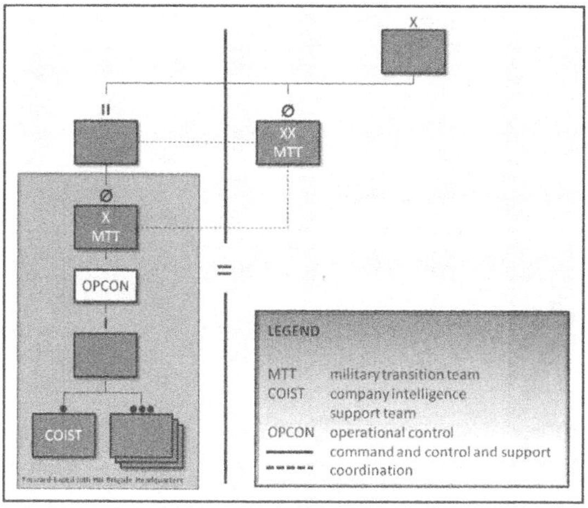

Figure 4-6. Military transition teams

SAMPLE TROOP-TO-TASK MODEL

4-32. The BCT may support assigned to multiple missions, including potentially supporting multiple foreign units in their operational area. Additionally, these FSF organizations may each report through different host-nation government channels and even to different ministries. To synchronize efforts in this case, U.S. forces must achieve unity of command and effort. Similarly, each of the FSF organizational commanders should synchronize their efforts with the host-nation government representatives, as appropriate.

4-33. Figure 4-7 depicts a sample troop-to-task model for a U.S. brigade. It depicts one company team acting as a quick reaction force; thus, the battalion retains the capability to conduct independent full spectrum operations. The figure depicts how U.S. company teams may form the basis of a brigade MTT or a brigade border transition team (and subordinate point of entry transition teams). The figure also depicts how a company team may provide police support. This support is in the form of two platoons supporting police transition teams and a third platoon in a joint security station providing support to a police transition team.

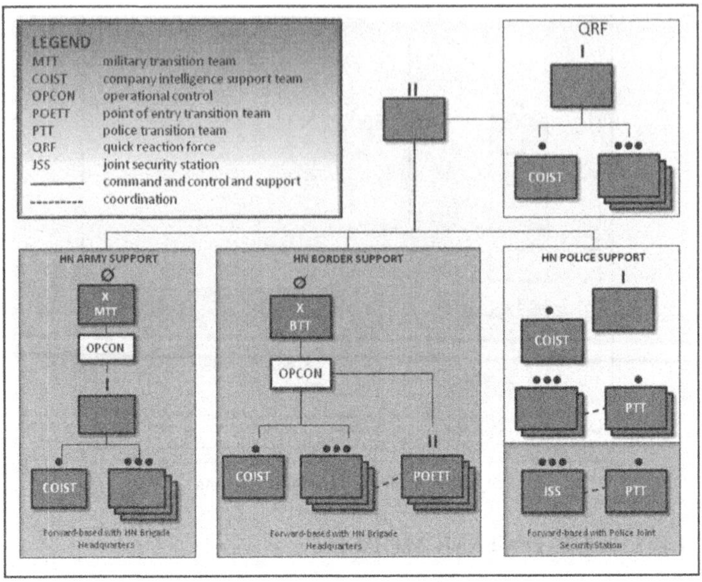

Figure 4-7. Battalions conducting security force assistance

4-34. Subordinate units of BCTs conducting SFA are best located inside a FSF base—not an American base. Co-location facilitates integration with the FSF. It allows the two forces to form mutual understanding and trust. Co-location and the close cooperation it facilitates often improves the population's perception of the legitimacy U.S. and foreign security forces, which can be an essential condition of the overall mission's end state.

4-35. If force protection requires, a U.S. area may be established in the FSF base, although this is not optimal. Key considerations for co-location may include the threat, FSF acceptance, physical space inside the FSF base, sustainment capabilities, medical facilities, and availability of quick reaction forces.

4-36. When U.S. forces are operating out of smaller outposts in an urban environment, the local populace sees the integration and presence of the U.S. and foreign security forces working together. This integration not only enhances overall operational effectiveness and trust, living and working together builds legitimacy of the two forces as well as FSF; it reinforces trust between the FSF and the people they are tasked to protect.

Chapter 5
Unit Operations

Unit operations require continuous activities. Commanders plan for premission, deployment, in-country preparation, employment, redeployment, and postdeployment activities.

PREMISSION

5-1. After receiving a mission, the unit continues detailed preparation activities. The unit obtains supplies, equipment, and training materials and prepares for deployment. Personnel prepare for and rehearse classes to be given in-country. They hone their military skills and conduct extensive area orientation. Paragraphs 5-2 through 5-10 discuss staff actions particular to security force assistance (SFA).

OPERATION PLAN

5-2. The S-3 disseminates the operation plan to unit staff and subordinates. The S-3 ensures predeployment training for Soldiers, to include preparation for training foreign security forces (FSF) and rehearsals for movement. The S-3 reviews the program of instruction for training FSF, to include getting approval from the commander and higher headquarters, if necessary.

CIVIL-MILITARY OPERATIONS

5-3. The civil-military operations section ensures the operation plan minimizes how operations affect the civilian population and addresses ways to mitigate the civilian impact on military operations. Civil-military operations also ensure the operation plan discusses coordination with any augmenting actors. These actors might include civil affairs battalions or brigades, provincial reconstruction teams, or United States Agency for International Development project officers in the operational area. During predeployment training, Soldiers receive training, materials, and briefings on the operational area. This training can cover the history, culture, religion, language, tribal affiliations, local politics, and cultural sensitivities as well as any significant nongovernmental organizations operating in the operational area.

TRAINING

5-4. As discussed in chapter 2, advisors should focus their premission training on the specific requirements of developing FSF. The training emphasizes the host-nation culture and language and provides cultural tips for developing a good rapport with foreign personnel. All premission training for SFA follows the training principles described in FM 7-0.

5-5. Based on the higher commander's training guidance, unit commanders assign missions and approve the draft mission-essential task list that supports SFA. The staff plans, conducts, and evaluates training to support this guidance and the approved mission-essential task list for SFA missions. Commanders prioritize tasks that need training. Since there is never enough time to train in every area, commanders focus on tasks essential for mission accomplishment.

5-6. Once commanders select tasks for training, the staff builds the training schedule and plans on these tasks. The staff provides the training requirements to the commander. After approving the list of tasks to be trained, the commander includes the tasks in the unit training schedule. The staff then coordinates the support and resource requirements with the S-3 and S-4. Finally, the commander ensures standards are enforced during training.

Evaluation

5-7. Evaluations can be either internal or external. Internal evaluations occur at all levels, and they must be inherent in all training. External evaluations are usually more formal and conducted by a headquarters two levels above the unit being evaluated. This subject must be carefully planned and discussed with FSF leaders to account for cultural sensitivities and current capabilities.

5-8. A critical weakness in training is the failure to evaluate each task every time it is executed. Every training exercise provides potential for evaluation feedback. Every evaluation is also a training session. Leaders and trainers must continually evaluate to optimize training. Evaluation must occur as training takes place. Emphasis is on direct, on-the-spot evaluations. However, leaders allow Soldiers to complete the task first. Leaders plan after action reviews at frequent, logical intervals during exercises. This technique allows the correction of shortcomings while they are still fresh in everyone's mind. The after action review eliminates reinforcing bad habits.

Specified Training

5-9. Augmentation elements require area orientation, refresher combat training, field training exercises, and the like. Unit training objectives are for developing capabilities to conduct internal defense and development (IDAD) activities for tactical operations, intelligence operations, psychological operations, populace and resources control operations, and civil affairs and advisory assistance operations in the host-nation language. Units identified for SFA begin intensified training immediately upon deployment notification.

5-10. After deployment to the host nation and before commitment to operations, the unit may receive in-country training at host-nation training centers or at designated training locations. This training helps personnel become psychologically and physically acclimated to the host nation. This training also allows commanders and staff some time to coordinate and plan within their own command and with civilian and military joint and multinational organizations. After commitment, training continues and is stressed between operations, using needed improvements identified in operations as the basis for training.

DEPLOYMENT

5-11. SFA is often conducted in semipermissive environments in which U.S. forces are guests of the host-nation or partner organization. Units move into the operational area by following their deployment standing operating procedures and the operation plan. Paragraphs 5-12 through 5-25 highlight organizations and procedures critical to SFA deployment. This information, however, does not obviate the need for extensive standing operating procedures and an operation plan.

PREDEPLOYMENT SITE SURVEY

5-12. The predeployment site survey unit aims to report accurately to its parent unit the existing FSF mission variables—mission, enemy, terrain and weather, troops and support available, time available, civil considerations (METT-TC). See table 5-1. It also establishes in-country command and control, support, protection, and sustainment relationships for the follow-on unit's mission execution. The site survey coordinates the in-country reception of SFA forces. This coordination wards off most friction during the handoff. This coordination requires all units—outgoing and incoming—to apply a comprehensive approach at all levels of command. Appendix D of FM 3-05.202 discusses site survey procedures in detail.

Table 5-1. Advisor team predeployment site survey checklist

Tasks	*Characteristics*
Area of operations overview	• Map familiarization. • In sector orientation. • Areas of emphasis.
Graphic control measures	• Unit boundaries. • Routes. • Other United States and multinational support locations. • Maintenance support location. • Medical support locations. • Other points of interest.
United States and multinational unit interviews by specialty staff	• Responsibilities. • Battle rhythm. • Required reports. • Standing operating procedures, continuity book. • Stay-behind equipment.
Briefings on advised unit	• Foreign security forces task organization. • Foreign security forces key leadership. • Intelligence assessment. • Unit operational assessment.
Advised unit introductions	• Among commanders. • Among staffs.

Advance Party Deployment

5-13. An advance party deploys to the host-nation or partner organization area of operations after the unit completes its mission planning and the proper commander representative declares it deployable. The advance party conducts predeployment activities and coordinates necessary travel documentation and country clearances. The advance party reviews the rules of engagement and any status-of-forces agreements before deploying. If applicable, the proper staff provides an updated threat briefing to the unit. The S-4 section and support company help with out-loading in the unit area. The unit places all supplies and materials on pallets beforehand and prepares the load manifest.

5-14. Upon arrival, the advance party processes through customs and immediately notifies the higher headquarters of its arrival and status. Often the higher headquarters provides point of contacts to the advance party. These contacts help the advance party obtain the proper identification, documentation, and weapons permits.

5-15. The advance party establishes contact with all U.S. and foreign security forces tactical unit commanders and conducts final premission planning activities. The advance party procures secure working, storage, and living areas for the unit and reconnoiters and prepares the training sites. The advance party coordinates for special support equipment to unload heavy supplies and transport the supplies and personnel to the training site.

Main Body Deployment

5-16. The unit deploys after completing its planning and the appropriate commander declares the unit ready. The unit follows the same procedures performed by the advance party. Members of the advance party meet the main body on its arrival and brief the main body on any required changes to the operation order.

5-17. After arriving in-country, the unit processes through customs and immediately notifies the higher headquarters of its arrival and status. The unit obtains the proper identification, documentation and weapons permits from U.S. or other sources, if applicable. Often the security assistance organization provides to the unit the necessary point of contacts that expedite this process.

5-18. The unit off-loads equipment and personnel onto vehicles for movement to the training site, or it stores the equipment in a secure area until the processing is completed. FSF secure the equipment but the unit members remember the commander is ultimately accountable for the advisor team's physical and personal security. Supply personnel stay with the equipment and help to guard it.

IN-COUNTRY PREPARATION

5-19. Upon arrival, the commander and S-3 brief the higher headquarters on the planned execution of the mission and reconfirm the required command relationship. Local conditions may require the unit to confirm or establish its in-country and external command and control, support, and sustainment relationships from outside its operational area upon arrival. The unit establishes direct working relationships with its next higher in- or out-of-country supporting element to—

- Determine the limits of the available support and expected reaction time between the initiation of the support request and its fulfillment.
- Confirm or establish communications procedures between the supporting element and the unit, to include alternative and emergency procedures for command and control, all available support, and medical evacuation.

5-20. The unit establishes procedures to promote interagency cooperation and synchronization. The unit—

- Identifies the location of the concerned host-nation, U.S., or other agencies.
- Contacts the concerned agency to establish initial coordination.
- Exchanges information or intelligence.
- Confirms or establishes other coordination protocols as necessary.
- Incorporates the newly established or changed procedures into the plans for mission execution.

5-21. The unit immediately establishes operations security procedures to support its mission execution and identifies rally points incorporated into its defensive, evasion, and personnel recovery plans.

5-22. After receiving a detailed briefing and further guidance from the advance party, unit personnel continue to develop effective rapport with the FSF commander and counterparts. They also assess their working, storage, and living areas for security and verify the location of the training site, communications center, dispensary area, and FSF troop area. With the FSF commander, the unit commander—

- Establishes rapport.
- Conducts introductions in a businesslike, congenial manner.
- Briefs on the unit's mission, its capabilities, and the restrictions and limits imposed on the detachment by the higher U.S. commander.
- Ensures all unit personnel fully support FSF and firmly believes a joint U.S.-FSF effort will succeed. Requests counterpart linkup be made under the mutual supervision of the FSF commander and the unit commander.
- Ensures all current unit plans are tentative and that assistance is needed to finalize them.
- Deduces or solicits the actual estimate of unit capabilities and perceived advisory assistance and material requirements.
- Recommends the most desirable courses of action (COAs) while emphasizing how they satisfy present conditions, achieve the desired training, and meet advisory assistance goals.
- Informs the higher in-country U.S. commander of any significant changes in the unit's plan to assist FSF.

5-23. The S-2 supervises the dissemination of intelligence and other operationally pertinent information within the unit and, as applicable, to higher, lower, or adjacent units or agencies. The S-2 also—

- Monitors the implementation of the intelligence collection plans to include updating the commander's priority intelligence requirements (PIRs), conducting area assessment, and coordinating for additional intelligence support.
- Establishes liaison with FSF intelligence and security agencies (within the guidelines provided by applicable higher authority). Assesses the intelligence threat and resulting security requirements, including coordination with the S-3 on specific security and operations security measures.

5-24. Through the S-2, the commander's PIRs are based on the latest information available and requirements for additional PIRs that arise from modified estimates and plans. The S-2 also—

- Analyzes the foreign unit's status to finalize unit plans for advisory assistance. These plans can include task organization of unit with counterparts, staff functions for planning SFA, and advisory assistance for executing SFA.
- Explains analysis to counterparts and encourages them to help with—and participate in—analyzing, preparing, and briefing the analysis to the foreign unit commander.
- Prepares and briefs the plans for training and advisory assistance.
- Helps the foreign unit inspect the available facilities to identify deficiencies. If the unit finds deficiencies, the S-2 prepares estimates of COAs for the FSF commander to correct them.
- Supervises the preparation of the facilities with their counterparts and informs unit and FSF commanders on the status of the facilities.

5-25. The unit ensures its security is based on the present or anticipated threat. Some recommended actions the unit may take include—

- Hardening its positions based on available means and requirements to maintain low visibility.
- Maintaining unit internal guard system with at least one Soldier who is awake and knows the locations of all other unit personnel. The guard reacts to an emergency by following an internal alert plan and starting defensive actions.
- Maintaining communications with all subordinate unit personnel deployed outside the immediate area controlled by the main body.
- Establishing plans for immediate defensive actions in the event of an attack or a loss of rapport with hostile reaction.
- Discussing visible security measures with foreign counterparts to ensure understanding and to maintain effective rapport. Unit personnel do not divulge sensitive information for the sake of possible rapport benefits.
- Encouraging the foreign unit, through counterparts, to adopt additional security measures identified when analyzing the foreign unit's status and inspecting its facilities.
- Coordinating defensive measures with the foreign unit to develop a mutual defensive plan. Unit personnel obtain from the unit's present reaction and defensive plans for attack. They encourage the foreign unit to conduct mutual full-force rehearsals of defensive plans; if unsuccessful, the unit conducts internal rehearsals of the plans.

EMPLOYMENT

5-26. Once the unit has deployed and arrived in-country, it begins employment. Employment is the conduct of SFA to support the FSF commander. Employment occurs within the advisor team itself. It includes foreign counterparts in the mission planning (preparing the FSF for the mission itself) to increase the capability and capacity of their planning processes, as well as increase the probability of success.

PERSONNEL

5-27. During employment, individual unit personnel perform functions described in paragraphs 5-28 through 5-38.

Commander

5-28. Before the mission, the U.S. commander advises and assists the FSF commander. The latter issues planning guidance for planning the execution of the mission and clarifies commander's intent. The U.S. commander advises and assists the FSF unit commander in the command and control process of tactical operations. By accompanying the FSF unit commander when the mission is received from higher headquarters, the U.S. commander assists any subsequent missions. The U.S. commander monitors how FSF understand the commander's intent and all specified or implied tasks.

5-29. During execution of the mission, the U.S. commander helps the foreign unit commander provide command and control during operations. After monitoring the tactical situation, the U.S. commander recommends changes to the COAs to exploit the situation. After monitoring the flow of information, the U.S. commander recommends improvements to use intelligence collection assets and to keep subordinates reporting required information.

Staff

5-30. Before the mission, the advisor staff advises and assists the foreign counterparts in preparing estimated COAs for essential tasks. The staff helps write tentative plans based on the planning guidance and unit's standing operating procedures. These plans include primary, alternate, contingency, and emergency plans.

5-31. During execution, the staff helps foreign counterparts coordinate the execution of the tasks. They disseminate portions of their plans to other personnel, senior and subordinate staff sections, and supporting elements. They help notify higher, lower, or adjacent staff sections of modified estimates and plans. Finally, the staff—with the S-2 and the S-2 counterpart—helps update the commander's critical information requirements with the latest information and requirements that arise from operations.

Executive Officer

5-32. The executive officer performs the organizational analysis of the unit's coordinating staff sections to ensure efficiency during the planning process according to initial planning guidance. With the foreign counterpart, this officer directs foreign staff sections as they develop estimates and plans. The executive officer monitors the liaison and coordination with FSF higher headquarters, recommending changes to improve efficiency.

Adjutant

5-33. The adjutant provides advice, assists, and makes recommendations to the foreign counterpart for conducting human resources. This includes monitoring the maintenance of foreign unit strength, accountability of casualties, unit morale, and postal activities. This may also include concerns with the foreign pay system, leave procedures, and casualty pay procedures.

Intelligence Officer

5-34. The intelligence officer advises and assists the monitoring of FSF operations security to protect classified and sensitive material and operations and recommends improvements. By helping the foreign counterpart update the situation map, the intelligence officer helps to keep the map current. The intelligence officer recommends improvements to the standing operating procedures of the tactical operations center communications so the intelligence section receives situation reports. This officer helps the counterpart monitor the collection, evaluation, interpretation, and dissemination of information. The intelligence officer assists in the examination of captured insurgent documents and material. This officer helps gather and disseminate intelligence reports from available sources to ensure the exploitation of all assets. The intelligence officer helps the counterpart to brief and debrief patrols operating as a part of reconnaissance and surveillance activities. Finally, the intelligence officer assists, with the advisor operations officer, developing reconnaissance and surveillance plans.

Operations Officer

5-35. The operations officer helps the foreign counterpart to prepare tactical plans using estimates, predictions, and information. This officer monitors command and communications nets, assists in preparing all orders and plans, and helps to supervise the training and preparation for operations. Finally, the operations officer monitors the planning process and makes recommendations for consistency with IDAD goals.

Supply Officer

5-36. The supply officer advises and assists the foreign counterpart in maintaining equipment readiness; monitoring the support provided to the foreign unit, its subunits, and attachments; and recommending improvements. Finally, this officer helps to supervise the use of transportation assets.

Civil-Military Team

5-37. Upon deployment, civil-military teams advise the commander and staff on civil-military considerations and coordinate efforts of any civil affairs units supporting the unit. The civil-military team mentors the counterpart team on the supported foreign element staff on civil-military operations, the legal and moral obligations of military commanders to civilian populations under their control, and the importance of respecting human rights. The civil-military team may introduce the counterpart to relevant nongovernmental organizations, United States Agency for International Development project officers, and provincial reconstruction team staff.

Company-Level Advisors

5-38. Advisors sent to elements below battalion level help foreign counterparts to analyze the mission and commander's intent from higher headquarters. They assist FSF leaders restate the mission, conduct an initial risk assessment, identify a tentative decisive point, and define their own intent. They also assist their foreign counterparts to analyze the mission variables. From these variables, advisors help their foreign counterparts to develop a COA that meets the higher headquarters concept of operations and commander's intent. Finally, they advise and assist in the conduct of operations and the flow of information to the FSF higher commander.

TRAINING

5-39. Training FSF during SFA parallels the training of U.S. forces. Training follows the principles of training described in FM 7-0. This training tailors the objectives to account for cultural differences and focus on the FSF mission and problem set.

Training Assessment

5-40. The commander of the advising unit begins with a training assessment of training plans designed during predeployment and passed on from the preceding U.S. units conducted in coordination with the FSF commander. This assessment is important to both evaluate the foreign security force but also as an exercise to establish and enhance the working relationship between U.S. and foreign security forces. The training assessment should cover all aspects of leadership, training, sustainment, and professionalization. To support an assessment, the advising unit should analyze the following specific foreign unit considerations:

- The unit's mission and mission-essential task list and capability to execute them.
- Staff capabilities.
- Personnel and equipment authorization.
- Physical condition.

- Any past or present foreign influence on training and combat operations.
- Operational deficiencies identified during recent operations or exercises with U.S. personnel.
- Sustainment capabilities, to include sustainment training programs.
- Internal training programs and personnel.
- Training facilities.

5-41. The level of professionalism of FSF, both units and individuals, should also be assessed. Adhering to established rules of engagement, ethics that meet the established laws and regulations of the commanding authority, laws for land warfare, and human rights, promotions based on merit and the security forces' support of civilian leaders and political goals all fall within this assessment.

5-42. The advisor unit, working with FSF leaders, also evaluates current members of the security forces for past military skills and positions. Often military reorganizations arbitrarily shift personnel to fill vacancies outside their knowledge and experience.

Training Plan

5-43. After completing the training assessment, advisors analyze the prepared training plan and determine if changes are necessary. The training plan stresses the deficiencies identified in the training assessment. It also identifies those in the host nation able to help train FSF and to strengthen the legitimacy of the process. Finally, the training plan should consider the unit's eventual self-sustainment. As the foreign security force gains sufficient capacity and capabilities to perform independently, advisors transition from a leading role to a mentoring role.

5-44. The plan for training FSF uses a comprehensive approach, supports the IDAD strategy, and has many echelons. Working with other echelons and actors within the area of operations can provide support and expertise that enhance the training and operations process as well as limit redundancy.

Program of Instruction

5-45. In coordination with the FSF commander and training staff, the advising unit develops a program of instruction. This program incorporates all training objectives that satisfy the training requirements identified during the assessment. The training program must support these requirements. The FSF commander approves the program of instruction prior to execution. When executing the program of instruction, advisors adhere to a training schedule consistent with changes in the mission variables. Advisors ensure through their counterparts and the FSF commander that all personnel receive training. Foreign counterpart trainers rehearse all classes approved on the program of instruction.

Presentation of Instruction

5-46. To present the training material properly, trainers follow the lesson outlines approved in the program of instruction. All training clearly states the task, conditions, and standards desired during each lesson, ensuring the foreign students understand them. Trainers state all warning and safety instructions in the host-nation language. The training to reinforce the concepts includes demonstrations of the execution of each task, stressing the execution as a step-by-step process. Trainers monitor students' progress during instruction and practical exercises, correcting mistakes as they are made.

Training Methods

5-47. An effective method of training used by U.S. forces is the crawl-walk-run method of teaching individual tasks, battle drills, collective tasks, and field exercises. (See FM 7-0.) This method is employed to develop well-trained leaders and units. During all phases, the training must include the mission of the unit in the context of the higher unit's mission to assist with the practical application of the training. Identifying the higher commander's mission and intent, as well as the tasks and purposes of other units in the area, would also add context to the training. This method can also be expanded to include the role of other actors.

5-48. At a minimum, units continue individual training to improve and sustain individual task proficiency while training on collective tasks. Collective training requires interaction among individuals or organizations to perform tasks, actions, and activities that contribute to achieving mission-essential task proficiency. Collective training includes performing collective, individual, and leader tasks associated with each training objective, action, or activity.

Collective Training

5-49. Collective training starts at the squad level. Squad battle drills provide key building blocks that support operations. Trainers can link battle drills and collective tasks through a logical, tactical scenario in a situational training exercise. Although this exercise is mission-oriented, it results in more than mission proficiency. Battle drills and collective tasks support situational training exercises, while these exercises support operations. Advisors must understand the operational environment when training FSF; training incorporates how internal and external threats and civilians affect the environment. Flexibility in using U.S. doctrine in training enhances efforts to make training realistic. Trainers modify U.S. doctrine to fit the FSF level of expertise, their command and control, the tactical situation, and sustainment base. Often the structure and capabilities of FSF differ from that required by U.S. doctrine. When FSF counter an insurgency, these exercises emphasize interplay among psychological and tactical, populace and resources control, intelligence, and civil affairs operations. Similarly, scenarios for each tactical exercise must include interplay among the various possible aspects of IDAD.

Individual Training

5-50. Training individuals of FSF emphasizes physical and mental conditioning, tactical training, basic rifle marksmanship, first aid, combatives, and the operational environment. Individual training also includes general tactics and techniques of security operations and the motivation, operations, and objectives of internal and external threats. Tough and realistic training conditions troops to mentally and physically withstand the strain of continuous offensive operations.

5-51. Foreign personnel cross-train on all types of weapons, communications and other equipment, and skills particular to their unit. Personnel losses must never cause weapons, communications equipment, or essential skills to be lost due to a lack of fully trained replacement personnel.

Small-Unit Leader Training

5-52. SFA frequently entail rapidly changing circumstances; thus, FSF small-unit leaders must be able to plan and execute operations with little guidance. Therefore, trainers stress small-unit leadership training which can occur concurrently with individual training. Tools the trainer uses to train leaders are manuals, previously established training, tactical exercises without troops, battle simulations, and unit missions. This training develops aggressiveness, tactical proficiency, and initiative. Small-unit leader training should include combined arms technical training such as procedures for forward observer and close air support. Leadership training must also include land navigation in difficult terrain and under conditions of limited visibility. Mission readiness and the health and welfare of subordinates are also continuous parts of training.

After Action Review

5-53. In any training, the after action review provides the critical link between training and evaluation. It is a professional discussion that includes the training participants and focuses directly on the training goals. An after action review occurs after all collective foreign unit training. Effective after action reviews review training goals with the concerned FSF unit commander. Advisor personnel ask leading questions, surface important tactical lessons, explore alternative COAs, assist the retention teaching points, and keep the after action review positive.

5-54. The advisor encourages the FSF unit commander to review the training with the entire unit or key subordinate leaders. At this review, the FSF commander stresses how to strengthen the chain of command. During the after action review, advisors avoid criticizing or embarrassing the FSF unit commander or subordinates.

5-55. If possible, after action reviews occur during the field portion of the training when the unit assembles at logical stopping points. These reviews provide feedback from instructors and leaders that increases and reinforces learning. This feedback provides a richer database for key points. During the after action review, the senior evaluator draws information from unit leaders to form possible alternative COAs.

Assessment

5-56. In all operations, including SFA, success is generally measured for three periods: short-, mid-, and long-term. Success in SFA is normally defined within the context of these three periods. In the short-term period, FSF make steady progress in fighting threats, meeting political milestones, building democratic institutions, and standing up security forces. In the mid-term period, FSF lead fighting threats and provide security, have a functioning government, and work towards achieving economic potential. In the long-term period, FSF are peaceful, united, stable, and secure; integrated into the international community; and a full partner in international security concerns.

Measures of Effectiveness

5-57. A *measure of effectiveness* is a criterion used to assess changes in system behavior, capability, or operational environment that is tied to measuring the attainment of an end state, achievement of an objective, or creation of an effect (JP 3-0). A measure of effectiveness provides a benchmark against which the commander assesses progress toward accomplishing the mission.

5-58. The following are tools to assess the effects of operations in meeting expectations defined for the short, mid, and long term. Although all SFA is not focused on a host nation, these tools are based on the four IDAD pillars:

- **Balanced development** attempts to achieve national goals through political, social, and economic programs.
- **Security** includes all activities implemented to protect the populace from the threat and to provide a safe environment for national development.
- **Neutralization** is a political concept that encompasses the physical and psychological separation of the threatening elements from the population.
- **Mobilization** provides organized manpower and resources and includes all activities to motivate and organize popular support of the government.

Measures of Performance

5-59. A *measure of performance* is a criterion used to assess friendly actions that is tied to measuring task accomplishment (JP 3-0). This quantitative analysis determines whether the task or action was performed to standard. The standard may be derived from a procedure or may be time-sensitive. Examples of measures of performance are listed in the appropriate Army tasks in FM 7-15.

5-60. The advisor helps the foreign counterpart to monitor the current situation for unanticipated successes, failures, or enemy actions. As the commanders assess the progress of foreign security operations, they look for opportunities, threats, and acceptable progress. Throughout the operation, the commanders visualize, describe, and direct changes to the operation, feeding the assessments of progress or regression back into the planning process.

REDEPLOYMENT

5-61. Redeployment is the return of forces and materiel to the home or mobilization station. Before redeployment, a unit conducting SFA inventories all supplies and equipment (jointly, if possible) to be passed on to the foreign unit or the relieving U.S. or multinational unit. These actions ensure all items are accounted for and the custody chain for property and equipment is unbroken. The outgoing unit also—

- Prepares a final evaluation report showing its estimate of the foreign unit's capabilities and giving an opinion of its future employment.
- Discusses the foreign unit's performance with its commander. The unit submits a copy of the foreign unit's final evaluation to the next higher FSF commander.

- Passes custody of training schedules, lesson plans, foreign operational records, and the foreign unit's final evaluation to the foreign unit and relieving U.S. or coalition unit.
- Presents debriefings, after action reviews, and the foreign unit's final evaluation to the higher in-country U.S. commander.

5-62. The outgoing unit commander supervises the redeployment. This commander approves prepared redeployment plans or, as necessary, directs changes to the redeployment plans. The commander and operations officer assess and select alternatives or changes to the present redeployment COAs based on the intelligence officer's intelligence estimate. They develop and disseminate a fragmentary order for selected alternatives or changes.

5-63. Unit personnel prepare all accompanying supplies and equipment for shipment. They brief ground support personnel on equipment handling procedures, routes, convoy procedures, and actions to take if a terrorist or insurgent incident occurs. They load personnel and equipment for movement to the departure airfield or airport following the load plan in the unit's operation order. They maintain accountability for all their personnel, supplies, and equipment.

Mission Handoff Procedures

5-64. During long-term SFA, conditions determine the rotation of in-theater units. Time is not the only governing factor. Changes in the operational environment may require reshaping force packages as situations change. In addition, internal administrative concerns might prompt or support a commander's decision to rotate units. Regardless, mission handoff is necessary and defined as the process of passing an ongoing mission from one unit to another with no discernible loss of continuity.

5-65. The overall authority for the handoff and the subsequent transfer of authority lies with the commander ordering the change. The authority for determining the handoff process lies with the incoming commander assuming responsibility for the mission. This changeover process may affect conditions under which the mission will continue.

Redeployment Considerations

5-66. Although intended for a direct handoff between U.S. units, commanders must make specific considerations when making a handoff to a multinational force. The considerations include mission, enemy, terrain and weather, troops and support available, time available, and civil considerations. For units relieved of a function by a government agency, procedures typically entail longer handoff times and more complex coordination. However, the other areas of consideration still apply and may in fact be a greater issue for an agency. Outgoing units that have past, present, or future projects planned with agencies prepare to transfer these projects to responsible agents in the incoming unit.

Mission

5-67. The outgoing unit must prepare a detailed study of their unit's mission statement and detail the current and implied mission tasks. The mission may also require an incoming unit with additional skill sets, such as advanced special operations, near real-time connectivity, civil affairs functional specialists, or complex media production ability. Knowing the mission, commander's concept of the mission, commander's critical information requirement, PIRs, and information requirements helps the incoming unit understand the mission. After a complete in-depth study of the operational area, the outgoing unit should help the incoming staff complete the handoff in a manner that allows for continued, uninterrupted mission accomplishment. The changeover must not allow the adversary to gain any operational advantages.

Enemy

5-68. The outgoing unit must provide the latest available intelligence on all threats that affect the mission. This intelligence includes comprehensive data on terrorists and terrorist-related incidents over the previous several months. In addition to the normal intelligence provided to the incoming unit on a regular basis, the situation may call for a liaison from the outgoing unit. Operations security is critical to prevent the enemy from discovering the impending handoff and then exploiting the fluidity of the change and the concentration of U.S. or multinational forces.

5-69. The in-country unit provides continuous information updates to the incoming unit. PIRs and information requirements were established for the original mission along with operational, strategic, and tactical information. The incoming unit must become familiar with the ongoing and upcoming mission PIRs and information requirements along with their links to planned nonlethal effects.

Troops and Support Available

5-70. To the incoming unit, learning about the friendly forces is as important as knowing the enemy situation. The unit must be familiarized with the command and control structure with which it will work daily. The incoming unit must know all friendly units in adjacent operational environments. It also must be aware of joint and multinational units and the capabilities of their mission support base. The incoming unit must also be aware of other operations, units, and their capabilities.

5-71. Mission handoff is the time for the outgoing unit leadership to introduce the incoming unit leadership to leaders of FSF. This practice helps prevent breaks in intelligence networks already developed by the outgoing unit and reduces the time the incoming unit needs to build trust. The outgoing unit supports the incoming unit's plans to prepare for the transition in counterpart relations. However, potential or anticipated friction between a foreign security force and the incoming unit may cause the relief to occur more slowly. Therefore, incoming and outgoing units need time for overlap to allow for in-country, face-to-face contact with their counterparts before the mission handoff. If possible, outgoing unit members should provide biographical data on their counterparts to include photographs. This information allows unit members to become familiar with their counterparts before deployment and to determine which advisor techniques may need more emphasis. Execution of the mission must continue within the capabilities of the incoming unit, the foreign unit, and the available supporting assets.

Time

5-72. The depth and dispersion of units and the number of operations conducted determines the time required to exchange units. Ideally, an overlap occurs to allow the incoming unit to become familiar with the operational environment and to establish rapport with their foreign counterparts. However, the handoff must occur as quickly as possible. The longer it takes, the more vulnerable become personnel in the operational environment. However, both outgoing and incoming units should not sacrifice continued and uninterrupted execution of ongoing operations for speed. The incoming unit needs enough time to observe training techniques and procedures and to conduct debriefing on lessons learned. Outgoing unit personnel need sufficient time to put incoming unit personnel in contact with agency counterparts.

Civil Considerations

5-73. The outgoing unit provides an in-depth area study, giving close attention to local problems. Often sister units can provide general demographic data that can be expanded upon for unit-specific needs. Popular support for U.S. activities occurring in the operational environment may directly influence changes in the mission statement. The outgoing unit provides this critical information and describes all completed and progressing civic action projects. The incoming unit considers how the host-nation government functions and international civilian or government agencies are involved in or influence the situation in its operational environment.

POSTDEPLOYMENT

5-74. The unit commander debriefs to provide an overview of the mission and all relevant information subsets. Topics can range from military geography, political parties, and military forces to insurgents, security forces, and ongoing agency operations. Concerns specific to a foreign security force and relevant in one SFA operation may not be relevant in another operation. The commander should codify and debrief any information subset that may affect mission success.

Postmission Debriefing

5-75. Redeployment is not the end of the mission. Upon arrival at the redeployment location, the unit undergoes an extensive debriefing. The intelligence staff section officer typically organizes and conducts the debriefing. The intelligence officer coordinates with higher-level intelligence organizations to take part in the returning unit's debriefing, particularly if other organizations tasked the unit to obtain information. All deployed personnel, to include attachments, must be available for the debriefing.

Documentation

5-76. After the debriefing, the unit commander and staff prepare two documents—an after action review and a report of lessons learned.

5-77. The after action review states the who, what, when, where, and how of the operation. It is a permanent record of the major activities of the unit from receipt of mission to debriefing. As such, it is an extremely important template on which past missions may be compared and future missions planned. Within 48 hours of being debriefed, the unit normally submits an after action review through command channels to the higher command. The intelligence and operations officers at each echelon keep copies of unit after action reviews.

5-78. Shortly after completion of the after action review, or simultaneously with its submission, the unit submits a report of lessons learned. This report documents the commander and staff's reflection of the operation and recommendations for the future. Units often prepare the lessons according to the six warfighting functions: movement and maneuver, intelligence, fire support, sustainment, command and control, and protection. This method addresses what worked and did not work during the operation, why it did or did not work, and what changes or substitutions are needed for existing tactics, techniques, and procedures in the unit.

5-79. As a minimum, the lessons learned report is forwarded to appropriate Service- and joint-level lessons learned agencies to allow integration into subsequent unit training and leader education. These agencies might include the Center for Army Lessons Learned and Joint Center for Operational Analysis.

Chapter 6

Sustainment

A brigade combat team may be required to execute operations across the spectrum of conflict. Hence, providing security force assistance (SFA) may require the sustainment staff to analyze the mission to determine support requirements and available capabilities. Since brigade combat teams conducting SFA may be responsible for a large operational area, an accurate and thorough assessment of U.S. assets as well as the foreign security force's assets is critical to mission success. The sustainment planner understands that though the mission can change, methods by which logistics are distributed within a theater of operations do not. Army sustainment enables the commander to execute a mission and sustain the force through full spectrum operations. Sustainment assets are allocated based on those requirements; no standard arrangement fits all situations.

As discussed in chapter 2, one of the principles of SFA is to ensure long-term sustainment. This principle dictates that all policies, strategies, and programs must be durable, consistent, and sustainable by foreign security forces (FSF) and their government once U.S. forces have completed assistance. This can be accomplished by developing the foreign security force's capability and capacity for asset production, management, performance, and maintenance.

SUSTAINMENT IN FOREIGN SECURITY FORCES CAPACITY

6-1. Sustainment is the provision of the logistics, personnel services, and medical support necessary to maintain operations until mission accomplishment. The endurance of SFA forces is primarily a function of their sustainment. Sustainment determines the depth to which the forces can conduct decisive operations, allowing the commander to seize, retain, and exploit the initiative. Endurance is the ability to employ combat power anywhere for protracted periods. Endurance stems from the ability to generate, protect, and sustain a force. It involves anticipating requirements and making the most effective, efficient use of available resources. Sustainment also enables strategic and operational reach. SFA forces require strategic and operational reach to deploy and immediately conduct operations anywhere with little or no advanced notice. FM 4-0 discusses sustainment in detail.

CORE LOGISTIC CAPABILITIES

6-2. Similar to any operation, the core logistic capabilities in SFA provide a framework to facilitate integrated decisionmaking, synchronize and allocate resources, and optimize logistic processes. The challenges associated with support cut across all the core logistic capabilities: supply, maintenance, deployment and distribution, engineering, medical support, logistic services, and operational contract support.

Supply

6-3. Supply operations include identifying requirements; selecting supply sources; scheduling deliveries; receiving, verifying, and transferring products; and authorizing supplier payments. To provide responsive supply operations in SFA to both the advising organization and FSF, logistic planning requires a collaborative environment that includes operations planning, maintenance operations, and the distribution system.

Maintenance

6-4. In SFA, maintenance operations aim to reduce repair cycle times by replacing components, modules, and assemblies as far forward as possible as well as maximizing reliance on parts distribution, visibility, and replacement. Advisors and their foreign counterparts require maintenance capacity for their freedom of action. Force movement and maneuver depends on sufficient readiness levels of transportation equipment and weapons systems. In SFA, maintenance encompasses both intermediate and organizational levels and consists of shop maintenance supporting the supply chain, transportation assets, and weapon systems.

Deployment and Distribution

6-5. The deployment and distribution capability moves forces and logistic support on time by meeting the required delivery date and providing time-definite delivery. By sharing critical information, units create unity of effort among diverse distribution organizations. This unity of effort enhances foreign support efforts and provides end-to-end support to satisfy deployment execution and sustainment operations for advisors and host-nation forces.

Engineering

6-6. Engineering provides capabilities to maximize freedom of action. These include capabilities and activities that support the maneuver of land combat forces; modify, maintain, or protect the physical environment; and contribute to a clear understanding of the physical environment by providing geospatial information and services to commanders and staffs.

Medical Support

6-7. A component of Army health systems oversee operational management of the health service support and force health protection missions for training, predeployment, deployment, and postdeployment operations. Army health system includes all mission support services performed, provided, or arranged by Army Medical Department to support health service support and force health protection mission requirements for civilian and military joint and multinational forces. For more information on Army health systems, see FM 4-02.

Health Service Support

6-8. All support and services performed, provided, and arranged by Army Medical Department promote, improve, conserve, or restore the mental and physical well-being of Soldiers and personnel in other Services, agencies, and organizations as directed. This includes casualty care, medical evacuation, and medical logistics. The health service support mission is a function of the sustainment warfighting function.

Force Health Protection

6-9. Force health protection measures promote, improve, or conserve the mental and physical well-being of Soldiers. These measures enable a healthy and fit force, prevent injury and illness, and protect the force from health hazards. They include some prevention aspects of Army Medical Department functions, combat and operational stress control, dental services, and laboratory services. The force health protection mission is a function of the protection warfighting function.

Logistic Services

6-10. Logistic services comprise the support capabilities that collectively enable the United States to rapidly provide global sustainment for U.S. forces. Logistic services include many disparate activities that are highly scalable capabilities. Included in this area are food, water and ice, base camp, and hygiene services.

Contract Support

6-11. Operational contract support provides acquisition functions to obtain materials, services, and construction of facilities for U.S. and host-nation forces in the local operating areas. The Department of Defense increasingly relies on contractors to perform many functions and tasks and then provides technical oversight. The expeditionary nature of SFA creates an even greater need for this capability.

Principles of Sustainment

6-12. The principles of sustainment are critical to maintaining combat power, maintaining strategic and operational reach, and providing Army forces with endurance. The principles of sustainment are anticipation, responsiveness, simplicity, economy, survivability, continuity, improvisation, and integration. While these principles are independent, they are also interrelated when used in planning and executing operations. For more information on sustainment principles, see FM 4-0.

Missions

6-13. All brigade support battalions have the same general capability to manage distribution and sustainment operations. The brigade support battalion provides support within an assigned operational area. It is force tailored and task-organized and uses subordinate companies, platoons, and teams to perform specific functions. In the sustainment role, the brigade support battalion is primarily concerned with the continuous management and distribution of stocks and allocation of maintenance in the operational area. Such assistance provides operational reach to maneuver commanders.

6-14. The brigade support battalion contains the following sustainment capabilities:

- Provides supplies, field services, field maintenance, recovery, and field feeding for itself and its assigned subordinates.
- Plans and conducts base and base cluster self-defense; defends against level I threats, assists in destruction of level II threats, and escapes or evades against level III threats.
- Has ability to operate as part of an Army or joint force.
- Coordinates host-nation support.
- Has ability to deploy an advance party to support early entry operations.
- Provides opening, distribution, and sustainment management information and advice to commanders and staff within its operational area.
- Exercises technical supervision over operations for all assigned units.
- Provides limited materiel management for internal stocks—classes I, II, III, IV, V, VII, and IX and maintenance management of internal assets.
- Provides a liaison team to augment other headquarters as necessary.
- Manages and maintains the its property records.
- Maintains data in support of Army equipment status reporting database and Army equipment status reporting system.
- Appoints contracting officer representatives to oversee contractor performance, certify receipt of services, and act as liaisons between the requiring activity and the contracting officer.

6-15. A brigade combat team often has attached companies such as an engineering company, a military intelligence company, an antitank company, or a network support company. Similar to the brigade headquarters and headquarters company, these attached companies are supported by the brigade support battalion regardless of where they are located on the battlefield. If one or part of these companies is task-organized to a maneuver battalion, it remains under the support of the brigade support battalion.

CRITICAL CONSIDERATIONS IN BUILDING HOST-NATION SUSTAINMENT CAPACITY

6-16. Effective SFA planning begins with accurately assessing and understanding U.S. policy objectives, determining the end state, and identifying resources available. Some possible logistic planning considerations—

- Have a holistic perspective.
- Address the causes, not the symptoms.
- Understand the desired end state.
- Understand the level of FSF senior leader support.

6-17. Logistic planning during SFA uses a holistic perspective. Operational environments can range from peacetime cooperation to counterinsurgency, stabilization to reconstruction. SFA forces determine the foreign security force's logistic and reconstruction requirements based on their internal defense and development strategy; they develop a sustainment end state to support the plan. Rebuilding the host nation's infrastructure and industrial capacity not only provides long-term job opportunities, but also builds trust and legitimacy in FSF, the host-nation government, and partner organizations.

6-18. Logistic planning during SFA addresses the causes, not the symptoms. Operational logistic support problems are the symptoms, not the causes. They result from problems that exist in the supporting systems or in the society's industrial capacity and economic strength. Teaching unit-level inventory management alone will not result in any sustainment improvement.

6-19. Logistic planning during SFA understands the desired end state. Depending on the U.S. strategic interest in a host nation and its security forces, levels of engagement can vary significantly to achieve the desired end state. Sustainment capacity-building activities can be force tailored and scaled to assist in any of the sustainment capability areas.

6-20. Logistic planning during SFA understands the level of FSF senior leader support. These leaders develop and set the policies for resource allocation. They support the various development projects and skills enhancement assistance activities. These senior-level advisors must fully understand the problems identified and an assistance strategy developed by SFA planners. From the beginning, SFA planners work with their counterparts to decide the plans and strategy to promote the vision and ensure continued effort.

6-21. SFA depends on capable logistic support for maneuver and sustainment. Without an improvement in sustainment capacity, FSF cannot significantly improve regardless of how much effort is devoted to advising and training. When planning for sustainment capacity building, planners thoroughly understand the logistic systems to identify causes that lead to functional deficiencies. They must have a holistic perspective and understand the various goals and resources required to formulate an operational approach that effectively and efficiently achieves U.S. objectives.

ASSET MANAGEMENT AND PERFORMANCE

6-22. One tool at the disposal of the commander and planners during an evaluation of foreign infrastructure and industrial capacity is called an asset management plan. This plan, commonly employed in the industrial community, manages an organization's infrastructure and assets to deliver a standard of service. An asset management plan provides information to FSF so they can analyze the capacity, condition, and current and future requirements of their facilities and infrastructure. Physical development should include facilities, transportation networks, area development, utilities, and other infrastructure components.

6-23. A description of the perceived problems of current foreign facilities, or problems created by missing assets, is also important. The description includes how the advisor intends to address deficiencies in foreign facilities and infrastructure. Each organization needs to determine each facility's function and value to their organizational mission by analyzing the current asset performance, costs, benefits, and improvements.

CURRENT ASSET PERFORMANCE

6-24. Asset performance is the most difficult portion of the asset management plan process to capture while assessing conditions of facilities and infrastructure. While historical records may exist, combat operations and possible instability of the host-nation government likely render these records irrelevant. Remaining structures, when measured against U.S. construction standards (designed to protect occupants from man-made or natural disasters), often are deficient, even if untouched by hostilities. Most existing facilities require significant modification to be considered for future use. However, FSF may use them in the interim until new facilities are completed.

6-25. Asset performance may also use gap analysis. Gap analysis defines the present state, the desired end state, and the gap between them. Once a gap has been identified, the commander and staff can develop courses of action to rectify or bridge the gap. Gap analysis alone is not adequate for all problem sets, as goals may evolve and emerge during the course of problem solving. Most problems have many alternative solutions.

COSTS

6-26. Establishing the short, medium, and long-term costs for facilities and infrastructure, from construction to control, is a critical step in prioritizing which facilities get built or refurbished first, and which get support at all. Units develop a forward-looking cost-profile for operating, maintaining, refurbishing, and replacing assets. This profile explains to civilian and military officials the implications and continued costs to maintain assets. SFA requires a plan for U.S. forces to hand off assets to FSF.

6-27. In constructing these models, the ideal cost-profile extends to cover the life of the longest-lived asset in the system. By doing so, the profile estimates the whole-life cost as well as determines the average annual costs. Furthermore, this cost evaluation may need to be revisited and updated annually, forming the basis for an annual bid for funding.

BENEFITS

6-28. Assets should provide some measure of benefit to U.S. and foreign security forces that can be measured or explained during and after SFA. This usually involves translating asset costs and projected performance into a monetary figure. Other benefits may be social or environmental, which may be difficult to quantify in monetary terms. Regardless, some attempt to record all the relevant benefits is important; the more quantitative, the better.

6-29. A benefit evaluation is vital to demonstrating that the ongoing cost of an asset system is reasonable given the scale of the benefits. The expenditure incurred by asset management needs to be justified in some way.

IMPROVEMENTS

6-30. How assets of FSF perform may be improved by acquisition, enhancement, or some other means. An improvement may address the potential reduction of performance and its effect on the rest of the asset system. The improvements represent a potential change to the performance and are typically managed as a project. Such management comes with an appraisal of the additional expenditure, comparing different options and selecting a preferred option based on the commander's guidance and host-nation requirements.

OPERATIONAL CONTRACT SUPPORT

6-31. Contracting is a key source of support for deployed forces across the spectrum of conflict. Because of the importance and unique challenges of operational contract support, the commander and staff need to fully understand their role in planning for and managing contracted support in the operational area. This includes understanding the types of contracted support, contract management, and contract close out.

TYPES OF CONTRACT SUPPORT

6-32. Current doctrine describes three broad types of contracted support: theater of operations support, external support, and systems support.

Theater of Operations Support

6-33. Contracts for theater of operations support provide support to deployed forces under prearranged contracts or contracts awarded from the mission area. They do so by contracting officers under the command and control of the combat sustainment support battalion. Contractors for theater of operations support acquire goods, services, and minor construction support, usually from local commercial sources, to meet the immediate needs of operational commanders. Theater of operations support contracts are the type of contract typically associated with contingency contracting. Sustainment brigades often are the requiring activity for theater of operations support contract actions.

External Support

6-34. Contracts for external support provide a variety of support to deployed forces. External support contracts may be prearranged contracts or contracts awarded during the contingency. These contracts support the mission and may include a mix of personnel including U.S. citizens, third-country nationals, and host-nation subcontractor employees. The largest and most commonly used external support contract is the logistics civilian augmentation program. This Army program is commonly used to provide life support, transportation support, and other support functions to deployed Army and joint forces.

System Support

6-35. System support contracts are prearranged and include, but are not limited to, newly fielded weapons systems, command and control infrastructure, and communications equipment. System support contractors, made up mostly of U.S. citizens, provide support in garrison and may deploy with the force to both training and real-world operations. The unit does not normally have a significant role to play in planning for or coordinating system support contracts other than coordinating and executing support of system support contract-related personnel.

CONTRACT MANAGEMENT

6-36. Contract management requires nominating and tracking contracting officer representatives for every service contract. It will also nominate a receiving official for all supply contracts. To ensure contractors provide the service or item per the contract, they need a quality contracting officer representative and official support. SFA units must identify and coordinate funding requirements in advance with the supporting resource management staff element to secure the correct amount and type of contract funding associated with the mission.

CONTRACT CLOSE OUT

6-37. In long-term operations like SFA, the advising unit ensures direct coordination and transfer of information related to operational contract support before mission handoff. Additionally, when the advance party arrives in the operational area, a designated Soldier actively seeks out current information on local contract support capabilities, policies, and procedures. This individual coordinates the formal handover of existing contract management responsibilities from the redeploying unit. This Soldier must know when recurring service contracts end since it takes 30 to 60 days to obtain funding approval. If the unit waits until the contract is about to expire before requesting additional funds, the sustainment brigade could lose the contracted service until funds are available.

EXTERNAL AGENCIES SUPPORT MANAGEMENT

6-38. Some external agencies help support the management of FSF. External support comes from interagencies, the host nation, and multinational organizations.

INTERAGENCY SUPPORT

6-39. The Department of Defense performs both supported and supporting roles with other agencies. It often is the lead agency for logistic activities in SFA. When tasked to provide military support to civil agencies, Army forces perform a supporting role. Sustainment forces may be tasked to support the U.S. ambassador that, as the President's representative, serves as senior U.S. Government officer in the country, or may employ resources of other U.S. Government agencies or private firms. Whether supported or supporting, close coordination is the key to efficient and effective interagency operations. In the absence of an American Embassy, or during operations in austere environments or remote locations, the advising unit may be the only organization able to provide supplies, services (including force protection), and life support. More typically, during the early stages of the advising unit deployment, the unit may find the American Embassy to be the primary source of supplies, services, and life support.

6-40. Nongovernmental organizations do not operate in either the military or governmental hierarchy. Their relationship with an advisor organization is neither supported nor supporting. An associate or partnership relationship exists between military forces and engaged nongovernmental organizations, ideally focused on unity of effort. If formed, the focal point where U.S. forces provide coordinated support to nongovernmental organizations would be the civil-military operations center of a joint task force headquarters.

HOST-NATION SUPPORT

6-41. Host-nation support and local procurement may provide sustainment, operational support, and tactical support. Host-nation support agreements fulfilling the command requirements for support need to be prenegotiated. Such support arrangements must be integrated into the distribution plan and coordinated with joint, allied, and multinational partners to prevent competition for resources and ensure high priority requirements are met. Host-nation support may include functional or area support and may use host-nation facilities, government agencies, civilians, or military units. Preestablished arrangements for host-nation support can reduce the requirement for early deployment of U.S. assets and can offset requirements for early strategic lift by reducing requirements for moving resources to the theater of operations.

MULTINATIONAL SUPPORT

6-42. Multinational support may consist of support provided from one multinational partner to another. One or more of the following organizational and management options facilitates multinational support:

- National support elements provide national support.
- Individual acquisition and cross-servicing agreements provide limited support.
- A lead nation provides specific support to other contributing nation forces.
- A role-specialist nation provides a specific common supply item or service.
- A multinational integrated sustainment unit provides limited common supply and support.
- A multinational joint sustainment center manages common-user logistic support.

6-43. In all cases, the multinational force commander directs specific multinational support within the applicable laws and regulations of the host nation. When operating within a formal alliance, the advising unit executes support following applicable standardization agreements.

FACILITATING THE REDEPLOYMENT PROCESS

6-44. Sustainment assets facilitate the redeployment and the retrograde process through its command and control of the distribution system. Many of the procedures are used to deploy forces, draw pre-positioned stocks, conduct reception staging and onward movement, and distribute supplies within the theater of operations apply to the redeployment process. Two factors in particular complicate redeployment operations: supporting operations and meeting requirements.

6-45. The same elements that operate and manage the theater of operations distribution system during deployment and sustaining operations perform similar roles during redeployment. When redeployment,

deployment, and sustaining operations occur simultaneously, the theater sustainment command may find it necessary to rebalance its forces or change the missions of subordinate units to effectively support ongoing operations and redeployment.

6-46. Requirements vary widely depending on the nature and scale of redeployment operations, infrastructure, and other factors. For example, redeployment operations could range from personnel only to entire units and their equipment. Depending on the political and military strategy, unit rotations may occur while decisive operations continue unabated or during operational transitions. Key considerations include, but are not limited to, several characteristics. These characteristics include the size of the force redeploying and deploying; infrastructure requirements and limitations; security requirements; traffic circulation; staging areas; distribution system capacity; quantity of supplies and materials to be redistributed; the amount and kinds of materials that require disposal agricultural inspections; and establishing accountability of retrograde cargo. The challenge is effective coordination and synchronization, vertically and horizontally, to ensure responsive simultaneous support to not only ongoing distributed operations, but also redeployment.

Chapter 7

The Advisor

Advisors are the most prominent group of U.S. personnel that serve with foreign security forces (FSF). They live, work, and fight with their FSF. The relationship between advisors and FSF is vital. Advisors are not liaison officers, nor do they command FSF units.

ROLES OF THE ADVISOR

7-1. The military advisor has three roles involving different responsibilities. First and foremost, advisors are members of a U.S. military organization with a well-defined chain of command and familiar responsibilities. Second, advisors embed themselves with their counterparts. Third, advisors are interpreters and communicators between U.S. forces and their foreign counterparts.

7-2. As members of military organizations, advisors receive and execute the orders of superiors. These orders may conflict with the orders their counterparts receive. Among other duties, advisors must act unobtrusively, but nonetheless positively, as inspector general—often observing, evaluating, and reporting on the performance of counterparts and their assigned unit.

7-3. Secondly, advisors live, eat, and work with the officers and men of their host units. Often, advisors soon regard themselves as one of them. The sharing of common hardships and dangers forges potent emotional ties. The success and good name of their units become matters of prime and personal importance to the advisor.

7-4. Finally, advisors are interpreters and communicators between U.S. superiors and foreign counterparts. Advisors must introduce and explain one to the other; they help resolve the myriad of problems, misunderstandings, and suspicions which arise in any human organization, particularly when people of starkly different cultures approach difficult tasks together. Advisors with quick and easy access to influential counterparts can sometimes be the best possible means of communicating.

7-5. To be effective, advisors obviously must gain their counterparts' trust and confidence. This relationship, however, is only a prelude to the advisor's major objective: inspiring and influencing a counterpart to effective action. In pursuing this goal—constantly, relentlessly, and forcefully, yet patiently, persuasively, and diplomatically—advisors must recognize conditions which can benefit or handicap their cause.

CONSIDERATIONS OF THE ADVISOR

7-6. An advisor follows ten considerations when serving with FSF.

By, With, and Through

7-7. All planned operations of a combat advisory mission must be conducted by, with, and through the FSF. Not counting immediate action battle drill responses, the mark of an effective advisory effort is the amount of personal involvement the FSF take in their own operations. Civilians must see that they are secured by their own security forces to promote the legitimacy of the host nation and their capacity and to build trust and confidence. In a word, combat advisors are shadows whose presence is felt at all times.

EMPATHY LEADS TO COMPETENCY

7-8. Empathy can be defined as identification with and understanding of another's situation, feelings, and motives. This is tough for experienced U.S. leaders and often harder to explain, but it is the key to the success of an advisory mission. Truly understanding other human beings and where they come from allows for honest relationships to develop. In most cultures, the place to begin understanding another person's feelings and experiences is by understanding the other person's narrative. The narrative is a collective group's identity as an interpretation of both ancient history and recent collective experiences. Delving into the narrative, understanding it and how it affects people, is the beginning of empathy.

SUCCESS IS BUILT ON PERSONAL RELATIONSHIPS

7-9. Developing a sense of empathy allows advisors to begin nurturing relationships with those they are tasked to advise. No amount of resources and firepower can compensate for the lack of a relationship between advisors and their counterparts. It must be honest, genuine, and heartfelt. Mutual respect, trust, and understanding create success. Both parties rely on each other for mission accomplishment and often for survival. This relationship is likely to be challenged on numerous occasions; only one built on mutual trust can survive and ensure mission success. Furthermore, advisors learn to pick their battles. Camaraderie and rapport with FSF can vanish in an instant if one does not prioritize and define critical lines in the relationship.

ADVISORS ARE BETWEEN U.S. AND FOREIGN SECURITY FORCES

7-10. Living and fighting with FSF allows for the development of true bonds, empathy, and trust. These desired results have a price to pay. When addressing their own Army, advisors soon realize that they are also not one of them, increasing their level of frustration. Advisors are often alone, navigating between two military systems and two cultures, never quite fitting in with either one.

ADVISORS WILL NEVER WIN, NOR SHOULD THEY

7-11. Often advisors find it difficult to understand what victory looks like in this environment. Progress tends to occur at a glacial pace and cannot usually be tracked on a day-to-day basis. The advisor attaining a tactical objective does not achieve success; success is achieved by the foreign security force achieving the objective. Advisors find it difficult to satisfy their own units, and they never fully satisfy the demands of their FSF. They are figuratively and literally caught in the middle. Advising is the art of striving to make a win-win situation for all parties. The advisor and foreign security force relationship is one in which the advisor removes the obstacles in the way. Only a plan by FSF will succeed—and it will only succeed if it is, in fact, their plan.

ADVISORS ARE NOT COMMANDERS

7-12. Advisors do not command FSF personnel or units. Advisors provide advice, training, and access to coalition resources to FSF. They are not intended to lead FSF in combat; they are responsible for the command and control of their own team of advisors. However, they can and should influence the commander of the foreign unit.

ADVISORS ARE HONEST BROKERS

7-13. Advisors are advocates for FSF with conventional forces. They transparently assess capabilities and reveal limitations of foreign units to the FSF's and advisors' higher command. They also provide access to U.S. resources—from close air support to medical evacuation—to FSF they advise.

ADVISORS LIVE WITH SHADES OF GRAY

7-14. Black and white is not available to the advisor. Caught between two cultures, systems, and narratives, the advisor works within a gray area. Unlike operations in a conventional unit, many of the checks and balances are not present. Advisors often find themselves isolated with great autonomy and no supervision. These conditions present moral and ethical dilemmas daily. Advisors need to be comfortable in this environment. Often they lose sleep at night and question their actions and those of their FSF. Thus, this mission calls for Soldiers of enormous character, moral courage, and intellect.

Talent Is Everything, But Advisors Must Understand Rank

7-15. Military forces around the world approach rank and prestige in completely different ways. In some cultures, rank equals experience; in others, it is nobility or education. Often many militaries have rewarded loyalty with promotions, which results in rank-heavy armies. Some FSF recognize talent, while others recognize rank. Advisors likely advise counterparts much more senior in rank. They must understand that rank on the uniform is important to many armies.

Make Do

7-16. Advisors will never have everything they believe they need to succeed. However, their creativity often distinguishes between success and failure. Austerity and economy of force can fail to meet the prevalent expectations that FSF have of a U.S. unit in regards to resources. It is up to the advisor team to close the expectation gap by its creativity and imagination. Scrounging, bartering, and negotiating are daily activities of advisors. An enormous amount of energy must be devoted to these activities to achieve mission success and endear them to their counterparts.

PERSONALITY TRAITS OF THE ADVISOR

7-17. Not every Soldier is well suited to perform advisory functions; even those considered to be the best and most experienced have failed at being an advisor. Effective advisors are only the most capable individuals. Advisors are Soldiers known to take the initiative and who set the standards for others; however, they are also patient and personable enough to work effectively with FSF. Recognizing that not all Soldiers are capable of performing as advisors, leaders should immediately remove advisors who do not exhibit these qualities.

7-18. Because advisors operate in very subjective environments, it is difficult to establish objective criteria by which to assess potential advisors. However, research and experience indicate that several personality traits greatly enhance the advisor's ability to adapt and thrive in a foreign culture. These traits include—

- Tolerance for ambiguity.
- Realistic when setting goals and tasks.
- Open-mindedness.
- Ability to withhold judgment.
- Empathy.
- Communicativeness.
- Flexibility.
- Curiosity.
- Warmth in human relations.
- Motivation of self and others.
- Self-reliance.
- Strong sense of self.
- Tolerance for differences.
- Perceptiveness.
- Ability to accept and learn from failure.
- Sense of humor.

7-19. Of the traits listed above, no single trait is paramount. Advisor selection and training programs seek to develop an understanding of the contributions of each of them.

SKILLS OF THE ADVISOR

7-20. Advisor-specific skills fall into two subcategories: enabling skills and developing skills. Enabling skills build on the individual and collective skills of advisors. Developing skills are the main advisor tasks of teaching, coaching, and advising. Advisors use these skills to develop the capabilities of the foreign security force.

ENABLING SKILLS

7-21. Advisors use enabling skills to work individually and in a group. Enabling skills help advisors communicate across cultures, build rapport, influence, and negotiate. Successfully employing these enabling skills sets the conditions for advisors to move forward with their mission.

Individual Skills

7-22. Individual skills help advisors properly perform their mission. Initially advisors should be selected based on subject matter expertise and not just rank. However, tactical proficiency does not necessarily equate to proficiency as an advisor. Modern-day military advisors must be able to teach and advise FSF. To do this, they must be knowledgeable and proficient in procedures and able to impart this knowledge effectively. They may also be able to advise FSF in the areas of intelligence, communications, operations, and logistics. Some advisors may be required to call in U.S. supporting arms, principally artillery and air support, and coordinate both air and ground casualty evacuation. Finally, advisors should be proficient in basic combat skills. These common skills, warrior skills, and survival skills are often required by advisors. Since these skills are critical to an advisor due to the isolated and independent nature of the mission, advisors should refresh them during predeployment training.

Collective Skills

7-23. As a group of advisors form a team, they must practice their warrior tasks, organize to cover the functions associated with any team, and familiarize its members with their duties and responsibilities. As with individual advisor skills, some are refined before deploying and others are learned in country. Figure 7-1 depicts this graphically.

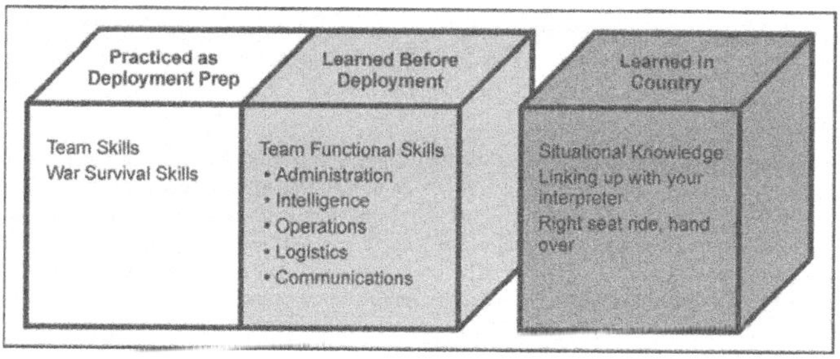

Figure 7-1. Collective advisor skills

DEVELOPING SKILLS

7-24. Developing skills—teaching, coaching, and advising—are required for every advisor. The advisor will make an initial assessment to determine the foreign unit's proficiency to determine the appropriate developing skill with which to start. Figure 7-2 illustrates the appropriate level of advice that corresponds to the foreign unit's capabilities.

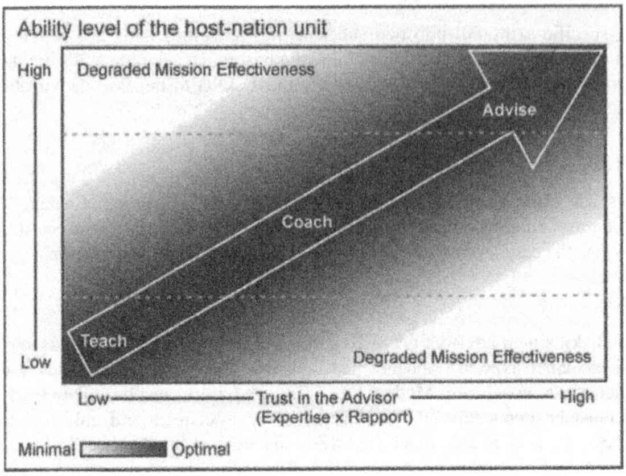

Figure 7-2. Appropriate advising level

7-25. Advisor teams have limited positional power and depend on personal influence to positively affect FSF efforts through teaching, coaching, and advising. Advisor team members often work with people of higher rank or grade than themselves. Advisors should remember the foreign unit's experience and capabilities and carefully choose opportunities to inject or impart knowledge. Foreign units are most receptive to advisor teams that teach unobtrusively. Foreign units most value those who are subtle in their teaching, coaching, and advising. Those advisors who master the ability to create a climate in which FSF personnel feel they are teaching themselves often prove the most effective.

Teaching

7-26. Teaching includes training and education. Methods of teaching can include classroom lectures, seminars, hands-on training, training exercises, and simulations.

Coaching

7-27. Coaching relies on guiding to bring out and enhance capabilities already present. Coaching refers to the function of helping someone through a set of tasks. Those being coached may or may not have appreciated their potential. The coach helps them understand their current level of performance and instructs them how to reach the next level of knowledge and skill. Coaching requires identifying short- and long-term goals and devising a plan to achieve those goals. The coach and the person being coached discuss strengths, weaknesses, and a course of action to sustain or improve the goals.

Advising

7-28. Advising is a combat multiplier that boosts supported unit capability. Advisors have experience in particular areas in which they are advising, but are not required to have similar backgrounds. Advisor relationships are not based on superior to subordinate relationships.

7-29. Advisors provide an expert opinion, advice, and counsel by focusing on both personal development (interpersonal and communication skills) and professional development (technical and tactical knowledge). Advising develops mutual trust and respect.

Advisor Versus Trainer

7-30. Advisors teach, coach, and advise work in a cycle. As FSF master one skill, the advisor can move on to other skills and initiate the process for the new skills. If FSF require additional teaching or coaching, the advisor can take them to that part of the cycle. Eventually, the teaching and coaching should decrease, and most of the advisor's time will be spent as an advisor providing an expert second opinion.

7-31. Ultimately, the question of what is the difference between being a trainer or an advisor will arise. Figure 7-3 shows that every advisor will have to be able to be a trainer, but all trainers are not expected to perform the more sophisticated task of advising. Hence, competent junior officers and enlisted Soldiers can be trainers, but only more experienced individuals can be advisors.

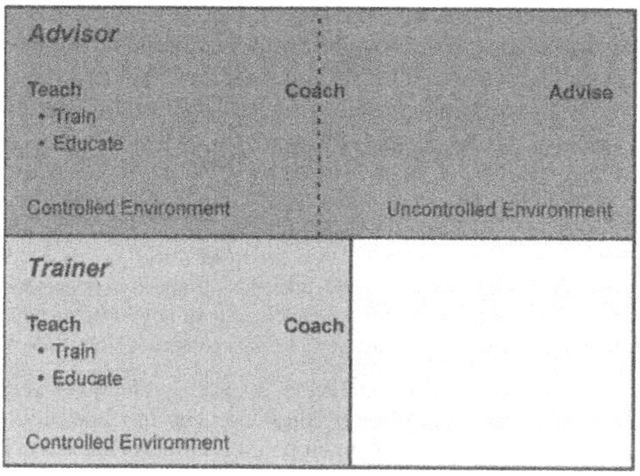

Figure 7-3. Advisor versus trainer

Chapter 8
Culture and Communication

As an advisor, understanding the host-nation population is a crucial element of premission planning and the development of foreign security forces (FSF). Prior knowledge of sociocultural differences aids in building effective relationships and prevents embarrassment, loss of rapport, and compromise of the mission. This chapter discusses three aspects of the sociocultural environment to provide a framework for analysis and comparison of foreign cultures for operational purposes. These aspects are society, culture, and communications.

SOCIETY

8-1. People who have a common culture, share a group identity, occupy a common territory, and are subject to the same political authority can be identified as a society. A society is not easily created or destroyed. Though they share the above factors, societies are not homogeneous nor static. A society usually has a dominant culture but can also have a vast number of secondary cultures, each influencing the other.

8-2. Advisors derive much of their effectiveness from their ability to understand and work with counterparts from a foreign society. Advisors must also consider societies outside their immediate operational environment whose actions, opinions, or political influence can affect the mission.

SOCIAL STRUCTURE

8-3. Society is composed of both social structure and culture. Social structure refers to the relations among groups of persons within a system of groups and is persistent over time. For the purposes of security force assistance, social structure includes groups, institutions, organizations, roles and statuses, and social norms. FM 3-24 contains a more in-depth discussion of social structure.

Groups

8-4. Within societies, groups are composed of two or more people that interact based on mutual interests and interrelated statuses and roles. These groups can be based on race, ethnicity, religion, or kinship. A racial group is defined by its members or by outside groups as different based on physical characteristics or descent. Race is a social category that has no biological basis. An ethnic group is a community whose learned cultural practices, language, history, ancestry, or religion distinguish them from others. Similar to racial groups, ethnic groups can be defined by their members or by outside groups. Religious groups may be subsets of larger ethnic groups and an ethnic group may contain members of different religions. Kinship groups are based on a link by blood relation. It may also be used to indicate a group of persons who live and work together to satisfy basic collective needs and goals. The role of a kinship group is generally seen to care for and educate offspring. This group provides for such needs as food, shelter, and clothing and prepares offspring to be functional members of the society in which they live.

Institutions

8-5. Institutions are the basic building blocks of societies. They are made up of groups that are organized to achieve a common task. Institutions are continuous through many generations and are built to continue to function even as individuals are replaced. An example of an institution from FM 3-24 is a military institution that brings together groups and individuals whose statuses and roles concern defense and security.

Organizations

8-6. Organizations are institutions that have a bounded membership with defined goals and established operations. They include fixed facilities from which they conduct business or operations and a set means of financial or logistic support. The four types of organizations are communicating, religious, economic, and social. Communicating organizations are designed to influence perceptions. Religious organizations are created to regulate societal norms, reaffirm beliefs and values, and provide social support. Economic organizations provide employment and assist in economic development. Social organizations provide support to groups in society, create social networks, and can influence ideologies. Organizations may belong to more than one category and exist to affect the behavior, values, and beliefs of society. As such, advisors must identify influential organizations both inside and outside of the FSF with which they are partnered.

Roles and Statuses

8-7. As members of a group, individuals interact based on assumed or given social positions. These social positions are referred to as statuses. Societies generally associate particular statuses with particular social groups (such as family, lineage, ethnicity, or religion). Each status inherently contains a set of expected behaviors known as a role, which includes how a person of that status should act, feel, and think. A status includes expectations about how others in a society should treat a person of that status.

Social Norms

8-8. Violation of a role derived from a given status results in social disapproval. The standard of conduct for social roles is a social norm. Norms may be either moral (incest prohibition, homicide prohibition) or customary (prayer before a meal, removing shoes before entering a house). Advisors must understand the roles, statuses, and social norms of groups they are advising. This knowledge will clarify the expected behavior of their FSF partners and their role as advisors.

LANGUAGE

8-9. Language is a learned element of culture that is a system of symbols that people use to communicate with one another. Successful communication within a society requires an understanding of the social setting, appropriate behaviors towards different statuses and roles, and nonverbal cues. The languages used in the advisors' operational environment have a major impact on the advising mission. Languages must be identified to facilitate language training during predeployment and employment, as well as in the requisitioning of translators. Translators are critical for communicating with FSF and interacting with local citizens and leaders.

CULTURE

8-10. Social structure comprises the relationships among groups, institutions, and individuals in a society; in contrast, culture provides meaning to individuals within the society. Social structure can be thought of as a skeleton with culture being the muscle on the bones. The two are mutually dependent and reinforcing. A change in one results in a change in the other.

8-11. Culture influences an individual's range of action and ideas, including what to do and not do, how to do or not do it, and whom to do it with or not to do it with. It is neither a random jumble of customs nor a laundry list of accepted practices and taboos that an advisor should know. It influences judgments about what is right and wrong and what is important and unimportant. Cultural rules are flexible in practice; the customs and actions that an individual practices are based largely on that particular environment.

CULTURAL CHARACTERISTICS

8-12. Culture is shaped by history, language, religion, and customs. Members of society use these shared beliefs, values, customs, and behaviors to cope with their world and with one another. It is learned behavior that is shared by members within a society. However, behaviors do not always define a culture.

Frequently, there are more telling cultural signs in the meaning attached to a behavior than in the actual behavior. Culture influences its members' identity, values, belief systems, and cultural forms. As such, it is arbitrary; advisors should make no assumptions regarding what a society considers right and wrong, good and bad. Culture is changeable through the social interactions between people and groups, both inside and outside of that society. TC 31-73 discusses cultural characteristics in great detail.

Identity

8-13. Each individual belongs to multiple groups through birth, assimilation, or achievement. Each group to which individuals belong influences their values and beliefs. Individuals consciously or unconsciously rank their identities into primary and secondary identities. Primary identities are those seen as core to their being; frequently these are national, racial, and religious. In contrast, secondary identities are more peripheral and may include such things as an individual's occupation or hobby.

Values

8-14. Values guide how individuals interact with the world around them. As deeply held beliefs about rightness, values shape how individuals think things should be. Values may never be articulated—in fact, they may not even be consciously held. At times, the beliefs of an individual may even contradict one another. Advisors should have some understanding of a counterpart's value system. Usually values are nonnegotiable. In working with a counterpart with conflicting values, advisors should persuade the individual that the advocated action or behavior really does not conflict with their values.

Belief Systems

8-15. Beliefs primarily take two forms: values (core) and opinions (peripheral). Values, or core beliefs, are not easily changed. Opinions, or peripheral beliefs, stem from thought and decisions. The totality of the identities, values, and beliefs that an individual holds compose that person's belief system.

Forms of Beliefs

8-16. Values are broad moral conclusions about the way to live life. These core beliefs are unstated, taken for granted, resistant to change, and not consciously considered. They reflect strong personal certainties drawn from faith, childhood environments, and life experiences. They lend shape and order to people's lives, helping them to decide if they settled on the right actions, friends, leaders, religions, and careers.

8-17. Opinions, or peripheral beliefs, are immediate conclusions about some aspect of the environment. They are usually quite specific and are open to debate, consciously considered, and easiest to change.

Types of Belief Systems

8-18. Belief systems involve cultural narratives and myths. The interpretation of these narratives and myths provide people with insight into how they should feel, think, and behave. The most prominent systems of beliefs tend to be those associated with religions and ideologies. However, any system of belief in which the interpretation of narratives impacts people's behavior (such as superstitions) can be a contributing component of a society's culture.

8-19. A belief system acts as a filter for new information. It is the lens through which people perceive the world. Beliefs of one group as to what is rational, normal, or true may appear strange, irrational, or illogical to outsiders. Understanding belief systems of various groups in an operational environment allows advisors to work with their foreign counterparts and influence civilians more effectively.

8-20. Belief systems play a fundamental role in the development of a society's values. Careful analysis of a region's belief systems is crucial to understanding the local value system. These systems influence religious doctrine, narratives, and myths and provide enormous insight into the culture of the community. Ultimately, the values which are most prized within a belief system are religious. These beliefs affect that culture's systems of social organization, learned behaviors, and relationship to the physical environment.

Religion as a Belief System

8-21. At its most basic level, religion is both the individual and communal expression of contact with the sacred. Notions of the sacred can vary widely in societies and may include deities and sacred items. In most cases, religions also involve symbols, rituals, and ethics or dogma. Religion can be both an individual practice and a group phenomenon.

8-22. Most societies have some variety of religion—typically a set of sacred beliefs and rituals that provide a common understanding of moral codes and proper conduct. Expressing a common religious understanding helps people to make sense of their world. Often religion provides explanations for human suffering, natural disasters, broken relationships, inequalities, and death. Some societies also have secular equivalents of religion. For instance, America is often remarked to have a civil or civic religion that can be seen in the images, symbols, ideas, and behavior of the political realm.

8-23. Because religion is such an integral part of culture, when advisors carefully prepare and analyze the mission, they must examine the religions and religious groups in the operational environment. Religious beliefs, leaders, and institutions are central to how many societies view the world. When planning any advising operation, planners consider how religion affects FSF and civilians.

Broad Impact of Belief Systems

8-24. In almost every society, followers of dominant religions have a strong societal impact. This influence may be exerted at varying degrees upon the individual citizen, sections of society, the economy, the military, and the political structure. TC 31-73 discusses these in detail.

The Importance of Understanding Culture

8-25. Advisors derive their effectiveness from their ability to understand and work with foreign counterparts from another culture. They must understand enough of their own culture and their counterpart's culture to accurately convey ideas, concepts, and purpose without causing counterproductive consequences.

8-26. Advisors must be aware of aspects of the local culture and history that influence behavior in their operational environment. Advisor team members must understand the reasons and motivations underlying personal interaction and must practice patience when working with their counterparts. Group norms guide individual behavior, and advisors must understand how individuals in a society tend to interact as members of a group, whether a race, ethnic, or kinship group.

8-27. Cultural understanding is not derived from demographic information provided to the military through country briefs prior to deployment. It is gained from studying, interacting, and understanding the people, religion, history, customs, and social and political structures within an area. For true understanding, it is necessary to live among the people, gradually understanding the subtleties and nuances of their culture.

Cultural Education

8-28. Culture influences a person's perception of the world. People from one society tend to view people from other societies through the prism of their own culture. Often, their perceptions are biased, are inaccurate, or involve assumptions of superiority. For example, Americans view their society and its culture very differently from the rest of the world. Likewise, as foreigners, advisors view local cultures through the prism of their own idealized culture. This creates a large gap between the culture the advisor observes and the host nation's real culture.

8-29. To narrow the gap of cultural understanding an advisor will encounter, a program of cultural learning should be undertaken. The end state of this cultural learning should be to gain the trust of foreign partners and civilians. Advisors will do so by demonstrating their understanding and sensitivity towards the host nation's culture and social structures.

8-30. An awareness of the cultural aspects of the operational environment can significantly enhance the effectiveness of an advisor. An extensive study program is best, but even a concise program is worthwhile if it provides advisors with the necessary cultural information. The best programs involve credible teachers who intimately understand the culture and have personal experience in the society. Training is dynamic and moves beyond the standard lecture and handout format. Small groups or a discussion panel work quite well. Cultural behavior and cultural familiarization handbooks can also be used. Since handbooks vary greatly in accuracy and quality, credible subject matter experts should be consulted to determine which documents to incorporate into the training. Table 8-1 discusses the training recommendations and tools.

Table 8-1. Cultural education recommendation

	Training Recommendations
Cultural Education	• Comparison of cultural values and social structures (United States compared to those of the operational area). • Local customs and traditions (for example, greetings do's and don'ts). • Geopolitical history (pre-colonial to contemporary and the orientation of each faction or party). • The role of religion in daily life.
	Training Tools
Resources	• Guest speakers native to the country of interest (for example, nongovernmental organization staff, foreign students, recent immigrants, or selected refugees). • Others who have worked in or studied the mission area (for example, special forces personnel, diplomats, and scholars). • Cultural familiarization handbooks.
Format	• Combination of briefings, small group discussions, and question and answer periods. • Handouts to augment—not replace—speakers. • Visual media, specifically slides and videos of the mission area.

8-31. Advisors with previous operational experience have addressed the importance of cultural education. Many civilian agencies have developed intercultural effectiveness programs designed for their personnel working overseas. Often these programs can be revised and tailored for military and paramilitary groups. If available, these programs are an excellent resource for predeployment preparation.

CULTURAL GAP

8-32. Each culture has its own societal rules regarding who a person may speak to, how and when the person may speak, and what topics the person may speak about. Many cultures rely on nonverbal signals to communicate. In such cultures, posture, expression, and actions convey more than spoken or written words. Advisors work to bridge the cultural gap by learning the language as well as nonverbal communication cues. To bridge the gap, advisors develop a sense of cultural awareness, recognize differences, learn to adapt, develop tolerance, and maintain personal contact.

Develop a Sense of Cultural Awareness

8-33. Advisors must recognize they are a product of their own culture. They must learn as much as possible about the culture of the people with whom they need to communicate. When communicating with people across cultures, advisors abandon any sentiments of ethnocentrism—the tendency of individuals to judge all other groups according to their own group's standards, behaviors, and customs. Such notions lead an individual to see other groups as inferior by comparison.

Recognize Differences

8-34. Each culture has its own way of accomplishing required daily tasks. Advisors must understand that each particular society may approach things differently; it does not mean that they are inefficient or less valid. Differences should not be seen as negative. Respect for counterparts must be maintained at all times.

Learn to Adapt

8-35. Advisors must be flexible and ready to adapt and adjust their behavior; however, they must be careful not to overdo their adjustment. Individuals who are overly flexible are often perceived as being insincere. The successful advisor must strive to act in a way that is appropriate to the target culture. Above all, advisors should be themselves and show sincerity.

Develop Tolerance

8-36. Advisors must develop a tolerance for deviations from accepted norms. Events or activities that may seem extraordinary to newcomers may be common practice in the culture. Advisors must be aware that members of the foreign culture may be astounded by that which is commonplace in the United States. Careful observation should be made before judgments are rendered about seemingly peculiar behavior.

Maintain Personal Contact

8-37. Personal contact is the most effective way to bridge organizational barriers. Interorganizational stumbling blocks are very real and the prejudices that arise from them are exacerbated by misunderstanding and ignorance. Often it is too easy to attribute negative attitudes and hostile motives to faceless groups. The advisor must employ superior interpersonal skills and work directly and closely with individual members of other organizations. This type of contact can effectively reinforce commonality and diminish the impact of disparity. These personal relationships are the key to effective interorganizational relationships.

Culture Shock and Adaptation

8-38. The term culture shock describes the anxiety experienced by persons in a completely new environment. It is not the result of a single event. Rather, the condition develops slowly from a series of minor events or conditions. Culture shock expresses the feelings of not having a set direction, not knowing what to do or how to do things in a new environment, and not knowing what is appropriate or inappropriate.

8-39. Culture shock occurs because the mind and body have to go through a period of psychological and physiological adjustment when individuals move from a familiar environment to an unfamiliar one. The cues received by all of the senses suddenly change. During the day the foreigner is bombarded with unfamiliar sights, sounds, smells, tastes, languages, gestures, rules, requirements, interactions, demands, systems, and expectations.

8-40. Some differences that advisors experience between their lives at their home station and their lives when deployed to a foreign location are obvious. These differences include language, climate, religion, food, educational system, and the absence of family and friends. Other differences may not be as obvious. These differences include how people make decisions, spend their leisure time, resolve conflicts, express their emotions, and use their hands, faces, and bodies to express meaning.

Stages of Shock

8-41. Five distinct stages occur in the culture shock process. These include enthusiasm, withdrawal, reemergence, achievement and reentry. Figure 8-1 depicts the stages of shock.

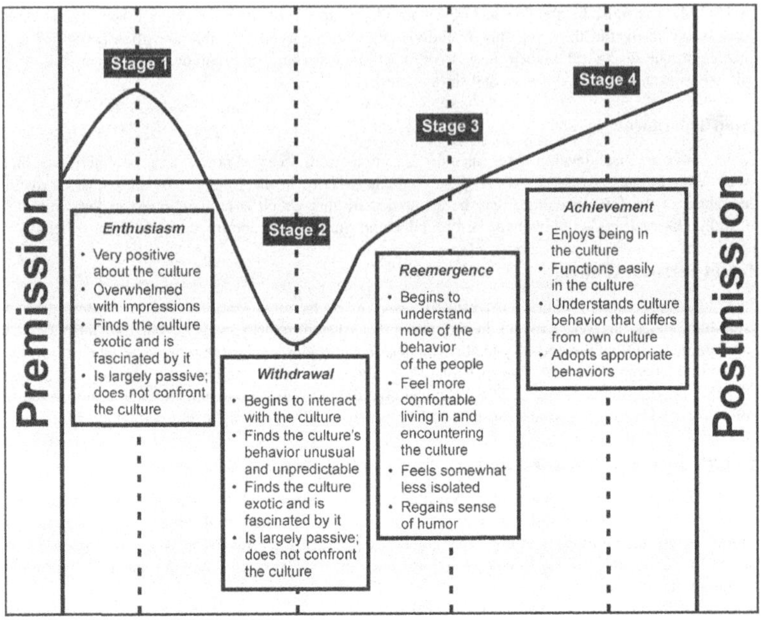

Figure 8-1. Stages of culture shock

Stage 1 – Enthusiasm

8-42. In this stage, a person may feel self-confident and pleasantly challenged. The person may be delighted by all of the new things encountered.

Stage 2 – Withdrawal

8-43. In this stage, a person begins to encounter difficulties and minor crises in daily life. It may be difficult to make oneself understood and foreign counterparts may prove more difficult to work with than anticipated. In this stage, one may have feelings of discontent, impatience, anger, sadness, and even incompetence.

Stage 3 – Reemergence

8-44. This stage is characterized by gaining some understanding of the new culture. A person may not feel as isolated, and a feeling of direction emerges. The individual is more familiar with the environment and is better able to adjust to the circumstances.

Stage 4 – Achievement

8-45. In this stage, the person realizes that the new culture has both good and bad things to offer. This stage is one of integration and the person is increasingly able to function in the new setting. A sense of accomplishment, a reduction of routine annoyances, and a more solid sense of belonging accompany this integration.

Stage 5 – Reentry

8-46. This stage occurs when people return to their own culture. They may find that things are not as they once were. Changes that occurred in their absence—their family, their friends, their communities, and themselves—combine to present a distorted image of home that differs greatly from the one imagined or remembered.

Culture Shock Symptoms

8-47. Being immersed in a society with extreme cultural differences tends to cause feelings of uncertainty and anxiety. The body and mind may react in unusual ways. Some persons may experience more pronounced physical symptoms of stress, such as chronic headaches or upset stomachs. Although uncomfortable, some degree of culture shock is a normal part of the adjustment process. Some common reactions include:

- Feeling irritable with (or even anger toward) one's own group or organization.
- Feeling isolated or alone.
- Tiring easily.
- Changing normal sleep patterns (either too much sleep or not enough).
- Suffering minor (but persistent) body pains, especially in the head, neck, back, and stomach.
- Experiencing feelings of hostility and contempt toward local people.
- Withdrawing from the local population (that is, spending excessive amounts of time alone reading or listening to music).

Overcoming Culture Shock

8-48. An important tool to overcome the obstacles of a new cultural environment is familiarity with the language. An ability to communicate in the new culture, even at the most basic level, reduces the effect and shortens the period needed for adjustment. Other means of overcoming culture shock can include having previous experience in the area (including a regular form of physical activity in personal routines) and acknowledging even slight progress in adjusting to the new culture.

8-49. TC 31-73 discusses culture shock in depth. It identifies the symptoms, cues, and treatment.

CROSS-CULTURAL COMMUNICATIONS

8-50. Communication is the transfer of messages from one person to another. These messages may be passed along verbally, in writing, or by signals (gestures, body language). The sender encodes the message and the receiver decodes it. The type and style of encoding used is based upon the sender's history, beliefs, values, opinions, and belief systems. Communication is determined by culture; as a result it can be either high context or low context. High context communication places an emphasis on clearly stating issues and problems and working towards resolving them in a straightforward manner. Low context communication places a premium on building relationships; as such, problem sets tend to be worked in a less formal matter in societies that utilize this type of communication. Quickly and effectively identifying whether a society uses high or low context communication is a key to effectively conveying material between the advisor and the host-nation population.

8-51. The receiver decodes messages based upon lifestyle, group membership, status and role, worldview, language, and social norms. Communication is a two-way process in which the encoding and decoding methods can affect both sending and receiving. Effective communication occurs when the message is perceived and responded to in the manner the sender intended. Ineffective communication can occur from poorly chosen words, flawed timing, a confused mixture of verbal and nonverbal signals, poor listening skills, and a failure to take culture into account. Communications can be broken into both levels and factors.

LEVELS OF COMMUNICATIONS

8-52. The cross-cultural communications capabilities required by an effective advisor to FSF can be described in three levels. At the lowest level is awareness followed by knowledge. The highest level is reached when these two are combined with well-trained and refined skills. TC 31-73 discusses communications in detail.

Awareness

8-53. Awareness is the basic level of cross-cultural capability. Awareness of cultural differences and their impact is the first prerequisite for successful work with a counterpart. Simply being sensitive to the fact that differences exist and carefully observing actions and reactions can assist the advisor in adjusting behavior and modifying actions to achieve greater influence with the counterpart. Awareness is not region specific and can be instilled in the advisor with relatively little training.

Knowledge

8-54. Knowledge of the details and nuances of a specific culture is the next level of cross-cultural capability. Advisors attain this level through academic study and immersion. Such knowledge is inherently area-specific and rarely transfers from one target area (or culture) to another. Developing in-depth regional knowledge necessary for effective cross-cultural communications requires extensive and time-consuming training. Appropriate personnel assignment policies must support this training.

Skills

8-55. Skills fundamental to effective cross-cultural communications, when combined with awareness and knowledge, form the highest level of cross-cultural capability. These skills include professional competence, language, nonverbal communication, negotiation, interpersonal skills, observation, problem solving, leadership, instruction, and fitness. Although some individuals show greater natural talent for these skills than others, all advisors require continual training in each to achieve and retain their full potential as advisors to foreign counterparts. These skills are discussed in more detail in TC 31-73.

FACTORS AFFECTING COMMUNICATION ACROSS CULTURES

8-56. There are four aspects that can impact communication across cultures. These factors are level of direct communication, perception of time, perception of the individual versus the group, and show of emotion.

Level of Direct Communication

8-57. Cultures fall along a spectrum of formality, with some cultures being direct and explicit in their communication and others less direct and more ambiguous. Cultures that rely on direct communication place more emphasis on verbal communications that are precise and explicit. Context and nonverbal cues are secondary to verbal communication. Cultures that rely on indirect communication rely less on verbal communication and more on context and nonverbal cues. Their verbal communication can seem vague and noncommittal to those from a culture that is more direct.

Perception of Time

8-58. Cultures can differ greatly in their perception of time. This difference divides into two cultural concepts of time: monochronic-time and polychronic-time. In monochronic-time cultures, members emphasize schedules, precise reckoning of time, and promptness. In such cultures, the schedule takes precedence over the interpersonal relation. In polychronic-time cultures, time is viewed as fluid. Members of polychronic-time societies do not observe strict schedules—agendas are subordinate to interpersonal relations.

Perception of the Individual Versus the Group

8-59. Cultures may be individualistic or collectivist in their orientation. An individualist culture is one in which the ties between individuals are loose—where people are expected to take care of themselves and their immediate families. In a collectivist culture, people are raised from birth into strong, cohesive groups. These groups offer a lifetime of protection in exchange for unquestioning loyalty. Advisors must understand from which culture their counterparts come to identify motivations accurately and communicate effectively.

Show of Emotion

8-60. Each culture has its own system of expressing emotion. Some cultures tend to express their emotions and show their feelings openly. Other cultures tend to repress those feelings more. Rather than showing their feelings openly, members of these cultures keep their emotions controlled and subdued. Those from the more expressive culture may view people from the repressive culture as cold or unfeeling. Similarly, those from the more repressive culture may view their more expressive colleagues as eccentric. To communicate effectively, advisors must be able to understand their foreign counterparts' emotional communication and display their own emotions carefully.

Chapter 9

Working with Counterparts

Rapport is a relationship marked by cooperation, conformity, harmony, or affinity. When people discuss good rapport, they describe a relationship founded on mutual trust, understanding, and respect. Relationships characterized by personal dislike, animosity, and other forms of friction often lack rapport. For the advisor, rapport describes the degree of effectiveness.

ESTABLISHMENT OF RAPPORT

9-1. Since advisors are in a unique military position, they establish rapport with their foreign counterparts. That position is one in which the advisor has no positional authority over the actions of their foreign counterparts. This lack of authority means that the doctrinal view of leadership is modified to emphasize interpersonal relationships and deemphasize authoritarian roles. Advisors instead use their people skills to influence the actions and decisions of their counterparts. Advisors use their people skills to building rapport. They have a genuine interest in other people, smile, remember and use people's names, encourage others to talk about themselves, listen to others, discuss what the other person is interested in, and make the other person feel important.

9-2. Advisors obtain certain knowledge before establishing effective rapport. First, they study FM 6-22 to gain basic leadership knowledge for understanding human nature and motivation. Advisors then incorporate information specific to the culture and society of their potential counterparts. This information may take the form of thorough area studies, operational area studies, and other research materials. Knowledge includes the components and techniques of rapport.

COMPONENTS OF RAPPORT

9-3. Three components compose rapport: understanding, respect, and trust. All the components of rapport are two-way streets; to have counterparts share about their culture, language, and experiences, advisors have to be willing and able to share also. Building this rapport may require the advisor to establish a personal level of understanding with their counterparts. Understanding is the first component of rapport. It begins before deployment and may include foreign cultural studies, language training, and equipment and doctrine familiarization. Once in country, advisors continue to broaden their understanding by observing and asking questions.

9-4. Respect is the next component of rapport. It is a reciprocal commodity. The foreign counterpart should grow to respect advisors, what they know, and how they perform. Advisors should look for those things that they respect in their counterparts. This may prove challenging. The counterpart may not fit the mold of U.S. or other coalition officers. Lacking formal training or education, the foreign counterpart may even be illiterate. The uniform standards of the host nation often may differ from the advisors' standards. Start with the basics and expand over time. Additional factors may affect the counterpart's willingness and motivation. For example, by accepting the duties, the counterparts (and their family) may be in mortal danger. The counterpart may have proven to be a fearless warrior, even without formal training. In short, traits deserving respect will exist even before the relationship matures. Mutual respect grows through shared experiences and shared dangers. Advisors should live, eat, and if authorized, fight with their counterparts. By sharing their hardships and dangers, advisors build respect.

9-5. Trust is the final critical component for building rapport. Trust grows gradually from understanding and respect. Building trust needs to begin on day one, but it will not mature until later in the relationship.

Advisors should begin by showing they are reliable and should do everything they say they will do. They should be in the right place at the right time. Two things can enhance this growth. First, the advisor should start out with confidence-building missions. Early success builds self-confidence and trust. Second, advisors should not promise any support they might not be able to deliver.

TECHNIQUES OF RAPPORT

9-6. Foreign security forces (FSF) respect advisors who actually show them the right way to do a task rather than just talk. Experience has shown that foreign units will not hesitate to go into danger if their advisors are right there with them. To further contribute to building rapport, advisors should maintain enthusiasm and a positive attitude; learn the language or how to use an interpreter; respect a counterpart's rank, age, status, and experience; develop negotiation skills; and improve interpersonal skills.

9-7. Other rapport-building techniques can include being confident, competent, and capable. Advisors never expect or demand their counterparts to do something that they are unable or unwilling to do. They demonstrate everything. Advisors learn their counterparts' names and spend time with them at meals and during holidays; they prepare mentally to interact with counterparts at all times. Advisors avoid creating an American-only enclave; they avoid giving the impression of favoring any one group. They frequently assess their counterparts' perceptions. Advisors recognize threats to discipline and enforce the chain of command.

ACCOMPLISHMENT OF RAPPORT

9-8. Advisors measure rapport by how well they can motivate their counterparts to act. The basic techniques of motivation (in the absence of authority) are advising, setting the example, seeking compromise, and coercing. TC 31-73 discusses rapport in detail.

ADVISING

9-9. Advising the counterpart to select a particular course of action is only effective if the counterpart trusts the advisor is professionally competent. If the counterpart does not perceive the proposed solution as realistic, the advisor's competence is questioned. Advisors carefully explain to their counterparts that recommended courses of action are realistic and will be effective.

SETTING THE EXAMPLE

9-10. Setting the example for the counterpart must be an ongoing effort for the advisor to avoid the appearance of a "do as I say, not as I do" attitude. In setting the example, advisors explain that what they are doing is the most effective action for the situation. This is particularly true when the counterpart really does not understand the behavior (or purpose).

SEEKING COMPROMISE

9-11. When seeking compromise with the foreign counterpart, advisors may create a situation in which the counterpart has a personal interest in successful execution. In some cultures, seeking a compromise may allow the counterpart to save face. Furthermore, in certain situations the counterpart—because of practical experience—may have a better solution to the problem at hand.

9-12. When seeking a compromise in certain cultures, the perceived competence of the advisors may suffer. This may be mitigated somewhat by approaching the compromise as two professionals (the advisor and the counterpart) reaching a mutual conclusion. To reach an effective compromise, advisors may have to conduct negotiations.

Note: Two areas of concern must never be compromised for the sake of maintaining rapport—operations security and human rights.

COERCING

9-13. Coercion is the least desirable method of motivation because it can cause irreparable damage to the relationship. Advisors should use coercion only in extreme circumstances. Advisors must avoid unintentionally forcing their counterparts into action. Instead, they establish and maintain good rapport by conveying they are sincerely interested in their counterparts, their nation, and its cause. Advisors will not belittle their counterparts' efforts. Advisors must demonstrate that they have come to help because they believe the counterparts' goals are just, fair, and deserving of success.

POLICIES AND AGREEMENTS

9-14. Advisors must remember their primary aim is to forward U.S. policy. Therefore, they establish relationships with foreign personnel, other U.S. forces, and other government agencies. Advisors promote U.S. policy through agreements, command relationships, and operations. TC 31-73 discusses policy in greater detail.

UNITED STATES AND HOST-NATION POLICIES

9-15. Advisors not only support U.S. national policy in the areas where they operate, they forward the policies of the host-nation government (unless otherwise directed by higher authority). Advisors maintain communications with higher headquarters to ensure activities align with U.S. objectives.

9-16. Host-nation policies may dictate practices and procedures that appear inefficient or uneconomical. Advisors should avoid criticizing or condemning such practices and procedures until they thoroughly understand them. Advisors should make private recommendations critical or contrary to host-nation policy.

AGREEMENTS

9-17. The status of advisors is normally specified by a status-of-forces agreement or other arrangement between the U.S. and host-nation governments. These agreements vary widely; they may offer little political protection or may provide for full diplomatic immunity. If no such agreement exists, advisors are subject to local laws and customs. Even if a status-of-forces agreement or other agreement exists that provides immunity, advisors are still expected to observe local laws.

COMMAND RELATIONSHIPS

9-18. Advisors must use the U.S. chain of command to obtain and disseminate guidance and assistance. They carefully distinguish between the U.S. and foreign security forces' chain of command. In particular, advisors prevent their counterparts from attempting to control subordinates through the U.S. chain of command. Advisors provide recommendations—not orders—to their counterparts. Only the counterpart should issue orders to subordinates.

9-19. Nongovernmental organizations and other government agencies may be co-located in the operational area. Advisors integrate the efforts of all organizations into their mission planning and should seek opportunities for integrated efforts with their counterparts. This type of unified action at the tactical level can be achieved through the planning process to integrate nonlethal actions. Holistically, nonlethal actions include development and diplomatic efforts.

CONDUCT OF OPERATIONS

9-20. Advisors can expect to participate in stability operations. These operations use military and paramilitary forces, as part of the interdepartmental team, to support internal defense and development operations. U.S. assistance may include advice on military organization, training, operations, doctrine, and materiel. In addition, the United States may provide and control U.S. logistics and sustainment for FSF. The objective of this assistance is to increase the capability of foreign organizations to perform their missions and operate efficiently in the given operational environment.

LEVELS OF ADVISORS

9-21. Organizations and individuals possessing greater skills and resources assist by imparting their knowledge through assistance efforts. The success of assistance depends largely on effective interaction between advisors and their foreign counterparts. Advisors may operate in the national, operational, and tactical levels. The national level is considered the largest national subdivision. This may be a nation's combined military headquarters, Service components, national-level government officials, or a combination. Operational-level advisors work at a level below the national-level advisor. This may be with military corps or division officers, with regional government officials, or a combination. Depending upon the size of the nation and its military, operational-level advisors may be assigned as low as brigades. Tactical-level advisors normally advise military units at or below the brigade level and provide counsel and assistance across the spectrum of conflict. They advise and assist counterparts in developing unit combat effectiveness and serve as liaisons between FSF and U.S. combat, support, sustainment and protection forces. Chapter 3 of TC 31-73 discusses these levels in detail.

CONSIDERATIONS OF RAPPORT

9-22. When building and maintaining rapport, advisors consider intelligence, human rights, role shock, ambiguity, personal relationships, bureaucracy, responsiveness, and diversity. These considerations are discussed in greater detail in TC 31-73.

INTELLIGENCE

9-23. To promote national policy and attain overall objectives, advisors may need to plan and conduct intelligence training. The advisor should assist the counterpart in developing a local intelligence collection program, training intelligence personnel in their respective specialties, and utilizing trained intelligence personnel properly.

9-24. The advisor should be familiar with area studies and area assessments. Pertinent documents should be compared to detect trends or changes. The advisor also should evaluate the foreign intelligence staff, its standing operating procedures, its chain of command, intelligence projects, and reference material available from other intelligence agencies. Advisors should prepare and maintain a list of essential elements of information.

HUMAN RIGHTS

9-25. Advisors stress the consequences of mistreating suspects, prisoners, or other persons taken into custody. These persons must be treated per Article 3 of the Third Geneva Convention. Article 3 requires care for the sick and wounded, protection of prisoners and detainees of all types from abuse or other harm. Murder, mutilation, and torture are expressly forbidden, as is humiliating or degrading treatment. Advisors must never be active participants in the conduct of such punishment. Furthermore, they must explain to their counterparts that they are obliged to report any atrocities of which they have knowledge.

ROLE SHOCK

9-26. Role shock results from the discrepancy between the roles individuals expect to play and the roles they actually do play. Role shock may also be a product of the tension created by individuals trying to do jobs themselves versus advising others how to do the jobs. Role shock tends to increase until about the anticipated midpoint of the tour and seldom disappears completely until redeployment. Symptoms of role shock resemble those produced by other stressful life events.

9-27. Many advisors become entangled in complexities of bureaucracy—both foreign and American—and become increasingly frustrated when seemingly familiar things fail to respond in expected ways. This continuing frustration can be a key contributor to role shock.

9-28. Although most advisors undergo some degree of role shock, over time they learn to take the personal, living, and social conditions in stride. More challenging are the problems that arise in connection with their jobs—their professional roles, relationships, personal achievement, self-development, self-determination, and self-image. Often advisors find their duties and activities during deployments differ somewhat from those expected. The actual duties and responsibilities are greater in scope, involve technical work outside of their assigned specialties, and require honed administrative (rather than military) skills.

AMBIGUITY

9-29. Most advisory assignments are ambiguous. What an advisor is expected to achieve is not always made clear. Furthermore, assignments may conflict with what they have trained to do, what others expect of them, or what they want or expect to do.

9-30. U.S. and foreign security forces may spend months—even years—negotiating a project. Still, the overall objectives or definition of the mission to be accomplished might be expressed in general terms. Various interpretations of what is to be accomplished arise. Even more views surface regarding the most appropriate means of reaching those goals.

RELATIONSHIPS

9-31. Ideally, foreign counterparts provide a working context for advisors and help them to understand and adjust to the new culture. Historically, those with good working relationships with their counterparts make better adjustments than those who have no counterparts or bad relations with their counterpart. Good working relationships may be difficult to establish and maintain. Most advisors approach their assignments as teachers and mentors. Foreign counterparts may resist such a relationship, particularly if they were expecting someone to take over and do all or part of the job. In addition, many counterparts want their advisors to share in at least some of the blame if something goes wrong.

9-32. The United States has little control over the availability or capability of counterparts. The assigned individual may be technically incompetent, lack professional commitment, have other interests, or be unavailable when needed. In some cases, numerous counterparts are assigned and rotate in their workings with the advisor. A single counterpart may not be assigned until the advisor has been on the scene for some time. In some scenarios, a specific counterpart is never assigned.

BUREAUCRACY

9-33. Although some Americans might disagree, the United States maintains a system that is—by world standards—relatively permissive and nonbureaucratic. As such, advisors may become aggravated when confronted with or obstructed by host-nation bureaucracy, red tape, administrative centralization, and a cumbersome decisionmaking process. Cultures that emphasize consensus in the decisionmaking process often prove quite frustrating for advisors. To be successful, advisors strive to understand the bureaucracy in which they operate.

9-34. Advisors should recognize the value of being or knowing the middleman. This role is sometimes one of the key contributions an outsider can make. Middlemen can move in, around, and out of organizations somewhat independently of the local hierarchy, protocol, and customs. They can link persons and offices that might otherwise have difficulty communicating. Many advisors dislike this role and refuse to practice it; however, the middleman can be a significant catalyst in any advisory operation.

RESPONSIVENESS

9-35. As frustrating as foreign bureaucracies may be, advisors often get even more irritated with the responsiveness of U.S. forces. Advisors frequently fail to account for distances, lead times, diplomatic issues, and other challenges when conducting security force assistance. Moreover, few Soldiers in the unit conducting security force assistance have had substantive experience in working in an interagency environment. These issues are further complicated by the usual tensions between headquarters and field staffs and between administrators and technical specialists. Expectations, values, and styles of behavior vary from group to group.

9-36. Deploying personnel may not be familiar with the tasks of development and the vast range of problems typical of underdeveloped areas. They come in eager to contribute their knowledge and skill, and anxious to make maximum progress in the limited span of their assignment. In so doing, they may fail to recognize that the factors that frustrate them—such as the lack of professional skills and facilities—are actually their reason for being there.

DIVERSITY

9-37. Organizations attract individuals who differ in abilities, knowledge, skills, work ethics, social behaviors, values, and attitudes. Advisors must be able to recognize personality types and develop techniques to cooperate effectively with each type.

RECOMMENDED PRACTICES IN BUILDING RAPPORT

9-38. Advisors may adopt many practices to help work with counterparts. Practices that help advisors build rapport are discussed in paragraphs 9-39 through 9-51.

ENSURE THAT THE ADVISOR PRESENCE IS UNDERSTOOD

9-39. Advisors entering an operational area work under the sponsorship of a military commander, mayor, village head, or some other recognized local leader. Prior to this, the advisor's arrival may need to be coordinated through the district, sector, or other appropriate local office. Clearances from distant national, state, province, or sector governments cannot compensate for a clear, local explanation of why an American advisor is in the area. This particularly applies to small, isolated communities where it is unusual for a stranger to appear for even an hour without being acknowledged and accepted by local leaders. Without explanations from locally respected persons, civilians often arrive at their own explanations, often to the detriment of the advisor effort.

FIND A BASIS FOR COMMON INTEREST WITH THE LOCAL PEOPLE

9-40. If advisors show appreciation for civilians as individuals, cultural gaps and language barriers can be overcome and common ground can usually be found. Advisors should listen and show interest when the locals speak. Initial conversations usually center on universal matters, such as food, shelter, clothing, health, and education. Over time, discussions may naturally move to the matter the advisor wants them to consider. Advisors are better received if they know something about earlier contributions in such matters as agriculture, folk art, religion, and architecture. Naturally, advisors prove more effective and appreciated if they can speak the language.

TRY TO UNDERSTAND WHY THINGS ARE DONE THE WAY THEY ARE

9-41. Although some local practices may seem strange, they generally have good reasons behind them. Advisors can discover these reasons with careful observation and a creative imagination. Food habits, family traditions, folk cures, and festive celebrations almost always have a great deal of human experience at their root. The advisor also needs to be aware that many villages contain rival subgroups and factions. This tension needs to be accounted for when working with the local people.

START WITH WHERE THE PEOPLE ARE AND WHAT THEY WANT

9-42. Advisors find out what the local people really want and work with them to achieve this aim. The need felt by the local people often is the best starting point. Once advisors address this aim, people become more appreciative and cooperative—they begin to raise their expectations and become interested in working for other improvements. To address the initial desires of the people, the advisor may need to call in personnel with the specialized skills needed for the particular project. Although this can result in a delay, it will help the advisor to achieve better cooperation.

Work Within the Local, Cultural Framework

9-43. Advisors understand basic cultural matters such as ethnic backgrounds of the people, family relationships, leadership patterns, and value systems. They also need some knowledge of local services such as health, education, communications, and transportation.

Help People Believe They Can Improve Their Situation

9-44. Most traditional peoples of South America, Asia, and Africa live in a largely static environment. Through their experience with change, they are more fearful of losing status than they are hopeful of bettering their condition. Therefore, changes suggested by advisors are often viewed with fear. Concrete local projects that yield easily observed benefits help convince villagers that they can improve their situation and make them more willing to cooperate in other projects.

Be Content with Small Beginnings

9-45. Change tends to come slowly in areas where there have been few in recent times. Advisors remember that knowledge—technical or otherwise—is cumulative. Once a small beginning has been made, greater activity and additional changes soon follow. It is easier to achieve momentum than to maintain it. Regardless, advisors start within as promising a framework as possible and with the support needed to sustain the momentum achieved.

Utilize Local Organizations and Recognize Their Leaders

9-46. The surest way for an activity to be continued after the advisor leaves is for it to have been launched and carried forward within the local organizational and leadership framework. People respond best when their local organizations are recognized as important and useful. Advisors consult and encourage recognized local leaders to contribute as they can. A well-conceived technical activity reflects credit on the local leaders associated with it. Advisors give attention not only to officials and family heads of local groups, but also to the quiet, behind-the-scenes leaders.

Help the Government Get Organized To Serve The People

9-47. For advisors to be most effective, they must understand the organization of the local government and how their activities fit into the overall scheme. Usually a set of agreements exists between various local agencies and the national government (usually through some sort of interministerial council). This agreement provides for a coordinated effort in servicing the varied needs of civilians. The advisor may need to work with appropriate agencies to assist in getting such agreements made. If such agreements already exist, advisors should be careful to recognize and strengthen them.

Train and Use Subprofessional, Multipurpose Local Workers

9-48. Selected young people in the villages can be trained and used as subprofessional, multipurpose village workers. Otherwise, advisors' influence is restricted to where they are standing and the immediate vicinity. Furthermore, advisors may spend so much time establishing and maintaining enough rapport with the villagers that they may become incapable of rendering any real service at all. Often the gap between civilians and the advisor is formidable because of their educational and cultural differences. Often the advisor works with villagers who are poor, illiterate, and devoid of outside contacts. Volunteer or paid local workers have proven quite helpful as they serve as a liaison between villagers and advisors.

Expect Slow Progress

9-49. As civilians begin to see successes from their joint efforts and begin to have new hope, they naturally want a larger hand in their own matters. The advisor may sometimes feel they want to assume more responsibility than they can carry. These evidences of growing pains should be appreciated, for they are a necessary part of becoming able to assume responsibility. The advisor needs to adjust to the growing desires of the people to help themselves.

TRANSFER CONTROLS CONSTRUCTIVELY

9-50. Advisors need to help civilians see how they can build the new—what they want—upon the foundations of the old—what they already have. From the beginning of a project, the advisor's team needs to talk with local leaders. Together they envision how to train local personnel, gather financial support, and transfer responsibilities before transferring control of activities. If operating responsibility is transferred too early, some breakage—usually of material things—will likely occur. If the team keeps control too long, the civilians who wanted to take over may grow disillusioned with them for not relinquishing control when they thought it should have been. Working with local leaders, the team determines when to phase out each technical activity.

EXPECT LITTLE GRATITUDE FROM THOSE HELPED

9-51. People who benefit from assistance sometimes feel defensive. In accepting assistance they are, in a sense, admitting their own insufficiency. The self-esteem of a person, community, or a nation as a whole is delicate. The advisor's team should not, therefore, expect thanks. Rather, the team should approach the people in a spirit of fraternity and humility, taking satisfaction in their forward progress. The team should do its job the best it can and accept work well done as is its own reward.

DYNAMICS OF ADVISING FOREIGN SECURITY FORCES

9-52. Advisors work in a complex environment that is heavily influenced by the unified effort of multiple organizations, relationships, compartmentalization, and organizational cultures. Paragraphs 9-53 through 9-57 discusses the probable dynamics that an advisor must take into account.

ADVISORS AND INTERDEPENDENT OPERATIONS

9-53. Advisors commonly operate in complex environments with interdependent operations. Interdependent operations require near-seamless integration of agencies representing the instruments of national power—diplomatic, informational, military, and economic. This same degree of integration must be extended to include external organizations and those agencies representing the instruments of power in partner states. The negotiation and cross-cultural skills that enhance their performance as advisors are also well suited to the complex interrelationships that characterize such operations.

ORGANIZATIONAL RELATIONSHIPS

9-54. Often organizational relationships can often be misleading and must be clarified. The actual interrelationships among and within organizations seldom follow a line-and-block diagram. Instead, they are heavily influenced by circumstances, personalities, perceptions, and resources. All relationships and lines of authority are subject to negotiation. Advisors understand that clarifying the roles, functions, and responsibilities is a continuous process and that each statement or action sets a precedent for future interrelationships.

9-55. The advisor must become hypersensitive to the most minor organizational disparities. The apparent similarities—the common language, sociological backgrounds, and citizenship—can be disarming and lead the advisor to overlook very real differences. If ignored, these differences can quickly impede synchronization.

COMPARTMENTALIZATION

9-56. Advisors balance security and synchronization. Overcompartmentalizing must be avoided. Often distrust can be prevented simply by not being overly secretive. Although advisors must conduct a careful, continual risk assessment to avoid compromise, the sharing of critical information is necessary for harmonious operations.

Organizational Culture

9-57. Advisors should approach external organizations as they would foreign counterparts—as unique institutions with distinctive cultures. Organizational differences, biases, and approaches must be researched and—whenever possible—accommodated. Each Service in the Department of Defense has unique perspectives that lead to significant differences in objectives and approaches to various problems. Profound differences in perspective may even be found between the different branches of the Army. If not recognized and accommodated, these characteristics may create distrust, disharmony, and disruption during critical operations.

Chapter 10

Cross-Cultural Influencing and Negotiating

Leveraging foreign counterpart cooperation can prove a difficult and critical task. Although coercion may prompt surrogate forces to perform a specific action, it can never lead to established conduct and legitimacy. Lasting advisors can achieve long-term success only by patiently applying their influence.

Advisors analyze their counterparts and determine their motivations and needs. Basic needs, culture, and individual personalities influence people. Advisors determine what these three things are so the appropriate influence techniques can be applied. Then advisors negotiate. Negotiation is fundamental to problem solving using the study, development, and honing of negotiation skills critical to the advisor.

INFLUENCING

10-1. Advisors need to gain influence with their counterparts, leaders, civilians, coalition members, and others beyond their chain of command. Advisors are self-aware, nonjudgmental, and capable. It is not necessary to accept others' beliefs or values, but advisors need to understand others' ways of thinking and operating. Knowing their own culture and understanding why they believe and value what they do helps advisors find a common cause. A common cause unites a diverse set of decisionmakers with whom advisors interact and try to influence.

10-2. Once advisors gain a minimum amount of influence, they can make things happen through others. They can extend their influence beyond their chain of command. Influence is simply getting others to do what you want.

10-3. Advisors identify the types and principles of influencing. They determine the best techniques with which to influence and review vital considerations. Using measures of effectiveness, they check that their influences have the desired results.

TYPES

10-4. Commitment, compliance, and resistance clarify how advisors understand the impact of their attempts to influence their counterparts. These three types of influencing are determined by how committed the foreign security forces (FSF) are to the task.

Commitment

10-5. Commitment is a willing dedication to a requirement or a cause because one wants to. Commitment-focused influence generally produces longer lasting and broader effects. Whereas compliance only changes immediate behavior, commitment reaches deeper—changing attitudes, beliefs, and behavior. For example, when advisors build desire and responsibility for a task among FSF, they often demonstrate more initiative, personal involvement, and creativity.

Compliance

10-6. Compliance is conforming to a requirement or demand because they feel like they have to avoid punishment or to gain a desired reward. Compliance-focused influence is based primarily on the leader's authority. Giving a direct order to a follower is one approach to obtain compliance during a task.

Compliance is appropriate for short-term, immediate requirements and for situations where little risk can be tolerated. Compliance techniques are also appropriate for leaders to use with others who are relatively unfamiliar with their tasks or unwilling or unable to commit fully to the request. If something needs to be done with little time for delay, and there is not a great need for the target of influence to understand why the request is made, then compliance is an acceptable approach. Compliance-focused influence is not particularly effective when a leader's greatest aim is to create initiative and high esteem with a counterpart.

Resistance

10-7. Resistance is a reaction against change. Resistance generally takes one of four forms—reactance, distrust, scrutiny, and inertia. It is manifested behaviorally (I won't do it), affectively (I don't like it), and cognitively (I don't believe it). Recognizing and lowering potential sources of resistance are key components to gaining compliance and commitment.

MOTIVATIONS

10-8. Different motivations influence behavior. One type of motivation is the desire for something. Another is the fear of something undesirable. As an agent of influence, advisors determine how to increase approach motivations and how to decrease avoidance motivations. For instance, recruiting for a special mission can use special badges as incentive (approach). Advisors can assure Soldiers that the mission is relatively safe by providing extra training and better equipment (lowering risks or avoidance).

PRINCIPLES

10-9. Six influencing principles are relevant to virtually any culture and any group. Understanding and applying these principles can increase an advisor's effectiveness in persuading a group or individual:

- Principle of scarcity.
- Principle of authority.
- Principle of social proof.
- Principle of liking.
- Principle of reciprocity.
- Principle of consistency.

10-10. With the principle of scarcity, people value more of what they can have less. They typically associate greater value with things that are rare, dwindling in availability, or difficult to acquire.

10-11. In the principle of authority, people are more easily persuaded by individuals perceived to be legitimate authorities or experts. They defer to experts who provide shortcuts to decisions requiring specialized information.

10-12. With the principle of social proof, people often look to the behavior of those around them for direction about what choices to make. This action is heightened when those around them are similar in terms of age, education, social standing, and experience.

10-13. In the principle of liking, people prefer to say yes and to comply with the requests of those they like.

10-14. In the principle of reciprocity, if someone grants favors, invited or uninvited, an overpowering need to repay that favor immediately blooms within the recipient of that favor. This human trait transcends all cultures and races.

10-15. With the principle of consistency, the desire for consistency motivates behavior. The drive to look and be consistent is a highly potent weapon of social influence, often causing people to act in ways clearly contrary to their own best interests. People do not like to appear inconsistent to others. It is human nature that people strive to feel good about themselves, which includes behaving per their important values and beliefs. When the behavior is consistent with whom people are and what they value, they feel good. People align with their clear commitments.

TECHNIQUES

10-16. The art of influencing involves knowing what techniques to use based on the situation. Advisors have to know the limits to which can use their influence; their influence is finite. They need to know when to use their influencing skills to achieve the highest payoffs. Their counterparts will only accept so much influencing if they don't completely trust their advisors. Table 10-1 lists positive and negative techniques delineated in FM 6-22. Table 10-2 lists resistance-based influence techniques. These are used when the counterpart resists the advisor's influences.

Table 10-1. Influencing and rapport techniques

Technique	*Description*	*Characteristics*	*Outcomes*
Pressure	Explicit demands, persistent reminders, sometimes appropriate in high stakes, urgent situations	Frequent checking Nagging	Gains compliance Frequent use may cause resentment, undermine morale, and harm relationship
Legitimate Request	Source of authority is used	Source credibility is important	Gains compliance
Exchange	Quid pro quo Reward offered for task completion; punishment threatened for incomplete task	Offer of reward or punishment not tied directly to getting job done, just something that counterpart wants or doesn't want	Gains compliance
Personal Appeal	Request based on friendship or loyalty	Agent of influence highlights talents of counterpart that make target the right person for the job (this is why you are the right person for the job, will you do it for me?)	Gains commitment Builds relationship
Collaboration	Works with target of influence by providing resources of some type, or by removing barriers to task completion	Offer of assistance is directly tied to request	Gains commitment Builds relationship
Rational Persuasion	Using facts, evidence, or logic in a persuasive message	Relies on agent being viewed as expert	Gains commitment Builds relationship
Apprising	Explaining the consequences of a potential action or decision	Decision is good for unit, career, or mission accomplishment	Gains commitment Builds relationship
Inspiration	Arousing strong emotion to build conviction or commitment	Appealing to values of the counterpart	Gains commitment Builds relationship
Participation	Requests for counterpart to be a part of planning, active participation in development of solution	Promotes commitment Possibly more feasible and culturally appropriate course of action	Gains commitment Builds relationship
Relationship Building	Build positive rapport and trust	Provide the foundation for using influence or inspiring	Gains commitment

Table 10-2. Resistance-based techniques

Technique	*Description*	*Characteristics*	*Outcomes*
Sidestep Resistance	Agent defines own role as consultant or advisor, but not commander Pushes the consequence of choice into future Depersonalize task to gain commitment to general idea Focus on end state instead of expended effort	Does not sell, just advises to prevent immediate resistance Puts less focus on immediate impact, more focus on long-term effect Stating "leaders should" instead of "you should" Stress mission accomplishment	Gains compliance Potentially gains commitment if paired with other influence techniques like rational persuasion
Address Resistance Directly	Address reluctance by lowering costs associated with tasks	Provide some of own resources through collaboration or counterargue concerns	Builds commitment
Address Resistance Indirectly	Build confidence, esteem, and self-efficacy for the task	Acknowledge difficulties while showing support for counterpart's abilities	Builds commitment
Distract Resistance	Interfere with the development of counterarguments	Distracting attention	Gains compliance
Disrupt Resistance	Disrupt complacency to bring attention to message	Shake up to motivate	Gains compliance
Consume Resistance	Resistance is energy	Wear them down prior to opportunities to resist. Persist with requests until they concede	Gains compliance
Use Resistance to Promote Change	Reverse psychology	Issuing a challenge	Gains compliance; risky when you don't know target of influence really well

10-17. To succeed with any influence technique, the influence attempt should be perceived as authentic and sincere. Positive influence comes from leaders who do what is right for Army forces, the mission, the team, and each individual Soldier. Negative influence—real and perceived—emanates from leaders who focus more on personal gain and lack self-awareness. False perception may trigger unintended side effects such as resentment of the leader and the deterioration of unit cohesion. With each and every influence attempt, Army leaders should ensure that the influence objective aligns with Army Values, ethics, the Uniform Code of Military Justice, the Warrior Ethos, and the Civilian Creed.

Considerations when Choosing Techniques for Influencing

10-18. Advisors clearly understand what their counterparts value and fear. This information allows advisors to quickly relate the objective of the influence to something that their counterparts treasure. Also, a strong sense of self supports the advisor's recognition of which techniques they can successfully execute. Someone with an intimidating presence may achieve compliance with pressure more quickly than someone without that presence. Understanding the context of the request is important. For instance, using participation in the midst of flying bullets would not make sense. However, a long-term project that impacts many people might be a worthy situation for participation. Advisors also consider—

- The motivation of their counterparts.
- The relationship they have with their counterparts.
- What they do or not have to offer.
- An understanding of the second- and third-order effects their recommendations will have.
- What has worked in the past.
- Possession of an appropriate power based for use of a technique in particular context.
- Consistency of technique with prevailing social norms and role expectations.
- Level of resistance encountered or anticipated.
- Costs of using a technique in relation to likely benefits.

10-19. The cost of a technique refers to the amount of effort, favor, or resources that accompany the technique. A personal appeal that relies on friendship is quite effective; however, counterparts might expect the favor to be returned. If resources or time are plentiful, then the exchange might not seem expensive, but if they are not plentiful, then advisors must choose a different technique. Advisors balance the costs and benefits of each technique.

Moderating Factors

10-20. Sometimes moderating factors diminish the impact of any influence technique. These factors may include cultural misunderstanding, level of trust, or constraints on the counterpart like threats or corruption. Advisors analyze and learn about these factors. Advisors must be flexible when using influencing techniques. When one technique doesn't work, switching strategies or adding techniques may be appropriate.

CONSIDERATIONS

10-21. Seven considerations need to be addressed to influence behavior. These include determining the goal; who to influence; motives; beliefs, values, and attitudes; cultural beliefs, values, and attitudes; and susceptibility. These techniques are designed to help advisors achieve their goals. Just knowing that the counterpart wants something is not enough. Whether it is time off, new weapons, or money, advisors have a limited supply of stuff they can promise; they can only bend so much on rules, regulations, and laws. It is at this point that the advisor enters the realm of negotiations.

Determine the Goal

10-22. Advisors influence others to achieve some purpose. To be successful at exerting influence, advisors have an end state or goal in mind. Sometimes the goal will be very specific. Many goals are less distinct and measurable but are still valid and meaningful.

Determine Who to Influence

10-23. Advisors develop a mental list of those whom they might need to influence. Getting a FSF platoon to take action may be as simple as influencing the commander, but the advisor may have to influence the soldiers and commander of the platoon. When practical, advisors consider many variations of whom to influence. They look for circumstances in the immediate environment that significantly affect their counterparts. Individual or group characteristics to consider can include—

- Gender.
- Religion.
- Age.
- Socioeconomic status.
- Ethnicity.
- Political affiliation.

- Level of education (important in determining how to access and persuade).
- Occupation.

10-24. The challenge is to determine which characteristics impact emotional or behavioral responses and under what circumstances. Ethnicity may significantly affect behavior and emotions regarding voting and politics but may have almost no effect on the decision to enlist in the military. Individual and group characteristics may provide additional insight on motivations.

Determine and Prioritze Motives

10-25. Motives are anything that is required or desired (needs and wants). Motives are the reasons behind the decisions people make. Motives come from an inner desire to meet a need or want. Advisors use the knowledge of what motivates others to influence them. Knowing one's counterpart and others who may be influenced gives advisors insight into influencing their behavior. The desires to fulfill, alleviate, or eliminate a need or want provides the motivation to change behavior.

Basic Motives

10-26. Basic motives involve physiological needs—food, water, and air—and safety needs—security and shelter. When people feel hunger, they are motivated to eat; when people feel pain, they are motivated to reduce the source of the pain. Such basic motives are extremely powerful in driving behavior and overwhelm psychological needs and wants.

Social Motives

10-27. Social motives associate access to other tangibles (money, goods, education, infrastructure, and healthcare) or more complex psychological motives with basic motives. For example, people learn to want money because it can be exchanged for food and other desired goods. Power, achievement, reassurance, escape, justice, and acceptance illustrate types of learned social motives. Examples include:

- Wanting better educational opportunities for their children.
- Wanting better paying jobs.
- Wanting their interests represented by the government.
- Wanting revenge for perceived wrongs.
- Wanting self rule.

Prioritize Motives

10-28. Advisors prioritize motives by immediacy of the need or want. Critical motives are immediate needs. Short-term motives are currently being satisfied or active efforts are being made to satisfy them in the near future. Long-term motives are desired but are not immediately important; satisfaction may be delayed until some point in the future.

Determine Beliefs, Values, and Attitudes

10-29. Although often difficult to derive, beliefs, values, and attitudes can prove quite effective in persuasion. Beliefs are what is thought to be true. People do not believe in something if they think it is not true. People vary from being gullible to skeptical. Values prioritize what one believes. Values can be ethical, doctrinal, social, aesthetic, or economic in nature. Attitudes reflect whether one likes or dislikes something. Likes and dislikes can often be traced back to a person's beliefs and values. This includes loves, hates, frustrations, and fears. Frustrations occur when a want is not met. By examining causes of frustration, the advisor may find another motive to address.

Compare Beliefs, Values, and Attitudes

10-30. The predominant cultural beliefs, values, and attitudes; traditions; and norms may differ greatly from the group the advisor is trying to influence. Many Americans act differently in different cultures. This change in behavior relates directly to their predominant culture.

Determine Susceptibility

10-31. Susceptibility is the likelihood that the individual or group will be open to persuasion. Determining susceptibility aids in prioritizing efforts. If susceptibility is moderate to high, then advisors may not need to focus as much effort as they would if the susceptibility is low. Advisors determine susceptibility by assessing the strength of the underlying source of the motivation and the perception of the desired corresponding behavior. The strength of the motivation depends on the perceived risk and the perceived reward, while the perception of the behavior directly links directly to how consistent the behavior is with the individual or groups' values and beliefs.

Perceived Risks and Rewards

10-32. Risks may include the threat of physical harm, financial loss, or disapproval of peers. Rewards may include financial gain, political power, approval of peers, or increased self-esteem. Risks and rewards can vary from immediate- to long-term. Immediate risks and rewards are stronger than long-term risks and rewards. For example, few people will walk a tightrope (immediate risk), but many people smoke (long-term risk). The higher the perceived rewards and the lower the perceived risks, the more likely the behavior will occur.

Consistency with Values and Beliefs

10-33. Advisors determine if the desired behavior is consistent with values and beliefs. Generally, behavior is consistent with the relative moral or ethical, religious, political, and cultural beliefs, values, and norms. Behavior that is inconsistent or incongruent with values and beliefs often decreases over time. Alternately, if behavior continues to be incongruent with values and beliefs, then individuals are likely to justify this behavior by modifying their values and beliefs. The more consistent a desired behavior change is with the individual's values and beliefs, the more likely the behavior will occur.

10-34. The lower the susceptibility, the less likely influencing behavior will be successful, which does not mean that the person cannot still be influenced. The lower the susceptibility, the more conditions will need to be modified for behavior change to occur.

APPLICATION

10-35. Influencing is a dynamic process. Advisors make a plan before attempting to influence. If one technique fails, they have to be able to shift to other techniques. Advisors do not need to sit down and write a plan before attempting to influence people. However, even with apparently positive results, advisors review their tactics and techniques.

CHECKS

10-36. Advisors check measures of effectiveness, unintended consequences, and second- and third-order effects. Advisors assess if their influencing is achieving their immediate measures of effectiveness. However, with every action some unintended consequences may occur or the activity may not prepare counterparts for more complicated tasks later on.

CROSS-CULTURAL NEGOTIATING

10-37. The principal form of negotiation is likely to be cross-cultural negotiation. Cross-cultural negotiation brings with it a series of unique challenges and techniques. Culture fundamentally affects language and behavior. It also significantly impacts the way people handle conflict.

10-38. Negotiations among members of the same culture can be stressful; negotiating with members of other cultures can be exceedingly difficult. Logically, working with other cultures is a basic skill for the Soldier and an absolute requirement when acting in an advisory role. Understanding the components, perspectives, roles, and outcomes will help minimize the difficulty of cross-cultural negotiation. Chapter 4 of TC 31-73 discusses cross-cultural negotiations.

Influences on Negotiations

10-39. Cultural differences may lead to a conflict between what the parties expect and what their families and communities expect. This has an obvious impact on negotiating behavior. Negotiations are influenced by perceptions, individualism, punctuality and pace, relationship building, and language.

Perceptions

10-40. Most Americans have certain preconceived ideas about people from other cultures. These perceptions may not be factually based, but they exist and influence negotiations. Similarly, foreign negotiators have certain perceptions about Americans. Again, these perceptions may not be based on fact, but they exist. Advisors need to understand how other cultures perceive Americans to appropriately adjust their negotiating style. Successful advisors find ways to capitalize on the positive preconceptions of their counterparts and find ways to neutralize the negative preconceptions.

Individualism

10-41. The United States encourages a largely individualistic culture. As such, the typical American negotiator prefers to enter into negotiations alone or, if required, as part of a very small negotiating team (two or three people, at most). This behavior is largely a by-product of U.S. culture, which focuses on the individual, performance, initiative, and accomplishments. Most other cultures of the world place less emphasis on the individual and more on the group.

Punctuality and Pace

10-42. Americans place a great emphasis on punctuality, viewing it as an indication of an individual's basic ability and commitment. However, other cultures emphasize punctuality differently. Some emphasize punctuality more than Americans. In other cultures, time is considered relative or may not be important at all.

Relationship Building

10-43. Americans are very competitive during negotiations and are inclined to stress short-term results. Interpersonal relationships are of minimal importance to American negotiators. Building long-term relationships only occurs after the successful completion of negotiations.

Language

10-44. U.S. negotiators must be considerate of others' with respect to language. Even if the counterpart is fluent in English, they may expect negotiators to speak their language or use an interpreter. Members of some cultures do not like to speak English as they are very proud of their language. When U.S. negotiator counterparts do not speak English, they will need an interpreter. Counterparts may also expect American negotiators to present detailed written proposals in the local language. Many cultures appreciate foreigners who take the trouble to learn a few simple courtesy words. However, some prefer that foreigners who are not fluent refrain from using more than a few, basic courtesy words.

Elements for Conducting Cross-Cultural Negotiations

10-45. The basic elements common to all negotiations must be applied in ways that allow for the differing cultures of the participants. Elements that must be considered include opening strategies, directness, strengthening behaviors, movement, power, face-saving, formal agreements, and mediation.

Components of Negotiating

10-46. Two opposite types of negotiating exist: distributive and integrative. They reflect a philosophy towards negotiating in general. Distributive negotiating is win-lose while integrative is win-win. Distributive negotiating involves dividing a fixed amount of a thing to the parties involved. This negotiation normally involves people who have never had a relationship before. Integrative negotiating involves joining two parties to achieve something together. Successful negotiations can involve a mix of both strategies.

10-47. When using distributive negotiating, advisors know the point at which counterparts will walk away. Advisors push them almost to that point by manipulation and withholding information. The distributive negotiator believes—

- The disputants are adversaries and the goal is victory.
- Fixed resources are divided so one side receives more than the other after making concessions.
- One person's interests or position opposes the other's interests or position.
- The dominant concern is usually maximizing one's own interests.

10-48. Integrative strategies used by advisors include cooperation, sharing information, and mutual problem solving. The integrative negotiator believes—

- The disputants are joint problem solvers seeking a wise decision.
- An understanding of needs will allow resources to be divided and both sides can win.
- The relationship between the parties is what is critical.
- The dominant concern is to maximize joint outcomes.

Perspectives Affecting Negotiations

10-49. Four perspectives impact negotiations: time, environment, self, and nonverbal communication. Advisors consider how each party views these conditions when negotiating.

Time

10-50. Some cultures and individuals see time as a micro view. It is all about here and now. Other cultures and individuals see time in a macro view. Now is a point that bridges the past with the future. For example, imagine a person with the first scheduled event has a deadline to make an agreement now. The next scheduled event is 1500 so that person has two hours to come to an agreement. The second person might agree that they can come to an agreement now; but to the second person, now could be within the next month or two.

Environment

10-51. Some cultures believe they can impact, if not control, their environment. If something should go wrong, somebody is to blame. Every accident is preventable. Other cultures believe the environment impacts the person. Accidents just happen and no one could have done anything about it. One side may be quite ready to commit to something in the future *knowing* that nothing stands in the way, while the other side may be reluctant to talk in certainties of the future *knowing* that factors beyond their control may prevent them from fulfilling their commitments.

Self

10-52. In some cultures, when a person makes a mistake, blame can be isolated to that person. In other cultures, the individual is seen as an element of the group. So a mistake is made not by the individual but rather the whole group. In negotiations, one might accept a personal slight to get the goal achieved, whereas the counterpart represents the group and cannot allow shame to fall on them even though one might accept it personally.

Nonverbal Communication

10-53. Since advisors will often have to work through interpreters, nonverbal communication becomes more important. Some cultures respect people who mean just what they say and maintain their composure. Other cultures respect the eloquence in which something is said or the emotion used to convey it. Where words may be translated improperly, the advisor or counterpart can hear tone and see emotion.

Roles of Negotiation

10-54. When entering a negotiation, advisors must know what their role. The three major roles associated with negotiation are negotiator, mediator, and arbitrator. Each role has its own tactics or techniques and separate functions.

Negotiator

10-55. Advisors most often find themselves in the role of a negotiator. A negotiation is the process by which parties exchange commitments or promises to resolve disagreements and reach a settlement. Advisors prepare for each negotiation so they will not be at a disadvantage during the negotiation. To prepare for negotiations advisors examine their options, discover what motivates the parties, list all positive scenarios, and dispute resolutions.

10-56. Advisors examine their options. They question their goals, the importance of the goals, and the amount of leverage they have. They determine who is in power and what to do if no agreement is reached. Advisors uncover what motivates their counterparts. They delve into their counterparts' underlying needs.

10-57. By listing all the scenarios that satisfy their needs, advisors can maximize the gain for all parties. Advisors can add features to the scenarios that increase the opportunities for all parties to be satisfied. When advisors dispute resolutions, they often find some form of compromise. One of the best outcomes of compromise is when all parties leave the bargaining table satisfied but no single party feels like they have won or lost the argument. A skilled negotiator should strive to achieve a win-win result.

10-58. The negotiator represents one side and has a vested interest in the outcome of the negotiation. It is best to avoid situations where there is a conflict of interests. By blindly pushing the coalition agenda, the advisor could lose the trust of the counterpart. Considerations for a negotiator should include:

- If the foreign security forces ask the advisor to negotiate for them, they expect the advisor, as their negotiator, to have their best interests in mind. The coalition may try to use the advisor to push their agenda and influence rather than negotiate.
- If the foreign counterparts take illegal action, advisors cannot use commitment techniques. They need to use some form of compliant technique.
- If the foreign counterpart is not violating any laws, the advisor should focus on commitment techniques. Success will not be achieved all at once.

Mediator

10-59. Advisors may find themselves in the role of a mediator. The mediator has three general roles to perform: facilitator, formulator, and manipulator. Table 10-3 identifies the roles and tactics advisors use.

Table 10-3. Mediator roles and tactics

Roles	*Tactics*
Facilitator	• Make contact. • Arrange for interaction. • Clarify the situation. • Supply missing information. • Transmit messages between parties. • Ensure all interests are discussed.
Formulator	• Control the pace, formality, and physical environment of the meetings. • Highlight common interests. • Help parties save face. • Keep process focused on the issues. • Make substitutive proposals. • Suggest concessions parties could make.
Manipulator	• Keep parties at the table. • Change parties expectations. • Highlight costs of nonagreement. • Reward concessions. • Promise resources. • Threaten the withdrawal of resources. • Offer to verify compliance. • Threaten to withdraw mediation.

Arbitrator

10-60. The foreign security force may try to put the advisor in the position of an arbitrator. An arbitration is the process by which the parties to a dispute submit their differences to the judgment of an impartial person or group selected by mutual consent or statutory provision. They want the advisor to decide between two positions. This is a dangerous position for the advisor. When a win-win solution can be found, the advisor can improve rapport with both parties, but when it is a win-lose situation, the advisor could alienate one or both groups.

Expected Outcomes of Negotiation

10-61. The outcome of negotiations sometimes can determine the outcome. Advisors consider the desired outcome to determine which confrontation preference to use. Negotiation can focuses on physical outcomes of relationships. How advisors combine their preferences to their focus affects their perceptions of the outcome.

Confrontation Preferences

10-62. Negotiators will have one of three preferences for confrontation: avoidance, collaboration, or competition. Those that prefer avoidance will consider accommodation as a viable means to resolve an issue. Negotiators that prefer collaboration consider compromise as the preferred outcome. Those negotiators who use competition want to win. Intentionally losing in a negotiation to avoid confrontation seems counterintuitive; however, it tends to protect the current relationship. Collaboration requires a mutual respect of the existing relationship, while competition is generally indifferent. It would be fairly easy to take advantage of a negotiator who prefers to avoid confrontation. At the same time, expecting to be able to compromise with a highly competitive negotiator may become a point of frustration. For the competitor, winning may be more important than the deal itself.

Negotiation Focus

10-63. Defining whether the negotiation is a win-win situation is not as simple as which side gained the most from the negotiation. Advisors often focus on physical outcomes whereas diplomats focus on relationships. Every negotiation varies on which factor is more important. Just as car dealers may accept less profit on individual services, they gain overall if the customer returns.

Perception of the Outcome

10-64. Comparing confrontation preferences leads to nine potential combinations. If advisors can determine a participant's preference—avoidance, collaboration, and competition—and focus, they can determine how the participant might feel based on the outcome of the negotiation. If one participant is a collaborator and one is a competitor, then a win-win situation is possible, but neither side will be completely satisfied with the outcome. The competitors will not have gotten a clear win, and the collaborators will not have gotten what they considered fair. Subsequently, they will both have some minor negative feelings toward the other.

Appendix A

Army Special Operations Forces Imperatives

This appendix outlines the imperatives for Army special operations forces conducting security force assistance and foreign internal defense operations. Table A-1 lists these imperatives.

Table A-1. Army special operations forces imperatives

Imperatives	*General Descriptors*	*Specific Actions*
Understand the Operational Environment	• Knowing the U.S. command relationships (military and interagency) • Knowing the population (who, what, when, where, why, and how) • Knowing the enemy and other antigovernment forces (crime, militias, gangs) • Knowing the higher commander's intent • Knowing that the press is a part of the operational environment (news reports of events reach the U.S. population quickly)	• Define relationship as partnership or augmentation • Conduct analysis using ASCOPE (areas, structures, capabilities, organizations, people, events) • Analyze the enemy order of battle • Protect the population
Recognize Political Implications	• Identifying the political (and legal) restrictions and implications of actions by U.S. and foreign security forces • Identifying political implications of establishing and enforcing rules of engagement policy • Identifying short and long term political implications of security force assistance • Understanding the importance of establishing human rights (critical for tactical and strategic success) • Acknowledging the effects of public opinion on the mission locally, in the United States, and internationally • Understanding that insurgents conduct operations to have political effects among the local populace, external actors, and the U.S. and multinational civilian public	• Include rules of engagement in the planning and briefs of operations • Include local government leadership in operations • Reinforce human rights briefings and training to all foreign security forces • Prepare all U.S. and foreign security forces to talk with the press • Prepare information engagements for all operations • Anticipate and neutralize the methods enemy forces pass propaganda

Imperatives	*General Descriptors*	*Specific Actions*
Facilitate Interagency Activities	• Conducting security force assistance as both a military and an interagency effort; taking advantage of the greater impact of collective efforts • Coordinating the scope and limitations of each agency's influence and programs • (as opposed to a purely military program) • Obtaining the active cooperation of the foreign security forces • Including military and civilian agencies working together for unity of effort	• Include interagency members in planning and training as appropriate • Provide assistance to support limited agency resources, allowing agencies to prioritize their primary roles • Conduct regular coordination meetings with civil agencies to synchronize security efforts • Include host-nation agencies whenever including U.S. agencies
Engage the Threat Discriminately	• Avoiding the use of overwhelming firepower—to prevent alienating the population and turning them away from the host-nation government • Giving training and advice to foreign security forces based on the mission and threat • Targeting insurgents, criminal activities (lawlessness), or subversion—to avoid alienating the populace • Minimizing the use of force in populace and resources control measures—weighed for potential gain versus potential cost • Ensuring that rules of engagement and escalation of force are balanced between force protection and safeguarding civilians	• Take civilians into account when considering weapons systems and techniques for operations • Include civilians in risk assessment during mission planning • Train and rehearse rules of engagement and escalation of force to prepare Soldiers to react in accordance to policy • Publicize reasons for use of populace and resource control (for example, curfews protect civilians); avoid enemy propaganda or rumors dictating the psychological response of the populace • Update rules of engagement and escalation of force periodically • Rules of engagement or escalation of force reflects engaging targets that are visually verified and not simply suppressive fire in populated areas
Consider Long-Term Effects	• Building up the foreign security forces and alleviating the root causes of the current situation over time—not just treating the symptoms • Solving problems in a broader political, military, social and economic context • Making policies, plans, and operations consistent with national and theater priorities and the objectives they support	• Understand local grievances and underlying issues • Interact with all aspects of the population a government to identify root causes • Ensure the commander's intent remains focused on higher goals and policy related to addressing root causes, and protecting the local population • Monitor and update rules of engagement as local government and police gain control of rule of law. Conduct operation that adhere to local laws as possible • Understand legal and political constraints (rules of engagement) to avoid strategic failure while achieving tactical success

Imperatives	*General Descriptors*	*Specific Actions*
Ensure Legitimacy of Special Operations	• Reinforcing and enhancing the legitimacy and credibility of the foreign security forces and government • Gaining trust and credibility to ensure advisors and U.S. forces obtain legitimacy with the local population • Obtaining legitimacy in theater, in the United States, and internationally to ensure ongoing support	• Conduct joint and multinational patrols and operations. • Ensure information engagement is common in all operations and contributes to maintaining credibility • Engage the media with the host nation to inform as well as to justify actions • Focus on what the population considers the root causes, not the U.S. perspective • Develop long-term solutions to restore or improve basic services with host-nation leadership
Anticipate and Control Psychological Effects	• Considering the psychological effects of all operations (such as populace and resources control measures, combat operations, and civic action) • Avoiding negative psychological effects that can negate tactical victories • Appreciating the power of people's perceptions versus reality	• All operations are developed, analyzed and executed with information engagement • Continue to gauge the psychological effects with the local population, local government, and security forces as well as the enemy (propaganda and interrogation)
Apply Capabilities Indirectly	• Teaching host-nation forces to plan instead of giving them tasks • Supporting U.S. efforts enhancing the legitimacy and credibility of the host-nation government and foreign security forces • Balancing control and pressure on foreign security forces to ensure they conduct successful tactical operations and accomplish missions • Fulfilling primary role to advise, train, and help foreign security forces take the lead and stabilize the host nation • Influencing a foreign security force without commanding it; not accepting a lack of effort or a high level of corruption	• Conduct training of foreign security forces only to teach basics • Transition to mentor or advisor relationship during planning and operations to encourage the host-nation to take the lead • Use discretion when advising counterparts as not to embarrass foreign security forces leadership • Use host-nation methods whenever possible

Imperatives	*General Descriptors*	*Specific Actions*
Develop Multiple Options	• Considering the complex and irregular environment • Planning to use the full range of capabilities • Anticipating that an operational environment may dictate a change of rules of engagement or type of operations • Ensuring foreign security forces are capable of shifting from one option to another before and during mission execution	• Continue to refine and train on battle drills and contingencies • Develop an understanding of skills and abilities of the members of the unit and community • Elicit the input of foreign security forces on how to address situations to widen the options beyond conventional techniques • Include government agencies, special operations forces, nongovernmental organizations, local police, and the host-nation government to develop additional methods or options for solving problems
Ensure Long-Term Sustainment	• Avoiding advising and training foreign security forces in techniques and procedures beyond their capabilities to sustain • Modifying tactics, techniques, and procedures; training; operations; and sustainment to fit the culture, education, and technological capability of the foreign security forces • Developing host-nation sustainment before handing over economic, social, political, military, and security initiatives and projects • Avoiding host-nation dependency on U.S.-funded programs	• Understand and utilize the equipment and manpower present in the foreign security forces unit's inventory • Assess the capabilities of foreign security forces for leadership, training, sustainment (logistics, administration, and equipment), and ongoing development of a professional security force • Recognize the programs that are durable, consistent, and sustainable by the host nation • Do not begin programs beyond the economic or technological capacity of the host nation
Provide Sufficient Intelligence	• Ensuring detailed, near real-time, all-source tactical intelligence products support all activities and programs • Ensuring detailed and comprehensive intelligence on all aspects of the operational environment and its internal dynamics support operations • Ensuring human intelligence, urban recon, and internal security elements assess internal threats, warn the government, and take action to penetrate and neutralize instability • Considering aspects of the society not directly related to the tactical combat situation when assessing the enemy threat • Using information from intelligence assets to advise, train, and help foreign security forces counterparts and ease interagency efforts	• Ensure host nation forces are conducting operations using accurate, real-time intelligence • Establish priority of effort when identifying intelligence requirements • Assist foreign security forces to prioritize and focus counterintelligence on the commander's critical information requirement to receive timely and specific intelligence • Assess other organizations that impede the rule of law and government functions (such as gangs, criminals, smugglers, militias, mercenaries, drug lords, and narcoterrorists)

Imperatives	*General Descriptors*	*Specific Actions*
Balance Security and Synchronization	• Maintaining unity of effort and avoiding compartmentalizing key security forces from planning and intelligence contingencies even when security concerns exist • Avoiding either excessive or insufficient security • Anticipating reasons foreign security forces may not provide intelligence, such as lack of internal security or lack of U.S. intelligence support	• Encourage combined planning, preparation, and operations with foreign security forces' military, police, and governmental units • Resolve conflicting demands on mission planning and execution • Employ measures to mitigate risk of operations security instead of dismissing the host-nation government, police, or military needed for the overall effort • Strive to gain the trust of the local population to gain the most human intelligence and support from the foreign security forces

Appendix B
Legal Considerations

GENERAL LEGAL BASIS

B-1. Law and policy govern the actions of U.S. forces in all military operations, including security force assistance. For U.S. forces to conduct operations, a legal basis must exist. This legal basis profoundly influences many aspects of an operation. It affects the rules of engagement, how U.S. forces organize and train foreign forces, the authority to spend funds to benefit the host nation, and the authority of U.S. forces to detain and interrogate. Under the Constitution of the United States, the President is the Commander in Chief of U.S. forces. Therefore, orders issued by the President or the Secretary of Defense to a combatant commander provide the starting point in determining the legal basis. Laws are legislation passed by Congress and signed into law by the President and treaties to which the United States is party. Policies are executive orders, departmental directives and regulations, and other authoritative statements issued by government officials. This appendix summarizes some of the laws and policies that bear upon U.S. military operations in support of security force assistance. This summary does not replace a consultation with the unit's supporting staff judge advocate.

LEGAL AUTHORITY FOR SECURITY FORCE ASSISTANCE

B-2. If the Secretary of State requests and the Secretary of Defense approves, U.S. forces can participate in security force assistance. The request and approval can go through standing statutory authorities in Title 22, U.S. Code. Title 22 contains the Foreign Assistance Act of 1961, the Arms Export Control Act of 1976, and other laws. It authorizes security assistance, developmental assistance, security force assistance and other forms of bilateral aid.

B-3. The request and approval for security force assistance might also occur under various provisions in Title 10, U.S. Code. Title 10 authorizes certain types of military-to-military contacts, exchanges, exercises, and limited forms of humanitarian and civic assistance in coordination with the U.S. ambassador to the host nation. In such situations, U.S. forces may be granted status as administrative and technical personnel based on a status-of-forces agreement or an exchange of letters with the host nation. This cooperation and assistance is limited to liaison, contacts, training, equipping, and providing defense articles and services. It does not include direct involvement in operations. Assistance to police by U.S. forces is permitted but, generally, Department of Defense (DOD) does not serve as the lead government department. Without receiving a deployment or execution order from the President or Secretary of Defense, U.S. forces may be authorized to make only limited contributions during operations that involve security assistance.

EXISTING UNITED STATES LAW

B-4. The Constitution divides the power to wage war between the executive and legislative branches of government. Under Article I, Congress holds the power to declare war; to raise and support armies; to provide and maintain a navy; and to make all laws necessary and proper for carrying out those responsibilities. Balancing that legislative empowerment, Article II vests the executive power in the President and makes the President the Commander in Chief of the Armed Forces. This bifurcation of the war powers created an area in which the coordinate political branches of government exercise concurrent authority over decisions relating to the use of Armed Forces overseas as an instrument of U.S. foreign policy.

B-5. The Supremacy Clause of the Constitution (Article VI) states, in part, that all treaties made by the United States are the supreme law of the land. Therefore, ratified treaties, such as the United Nations (U.N.) Charter (see paragraphs B-7 through B-8) and the Geneva Conventions, create legal obligations on U.S. forces regarding their ability to perform various types of missions and functions.

B-6. The War Powers Resolution of 1973 ensures the collective judgment of both the executive and legislative branches to deploy U.S. forces by requiring consultation of and reports to Congress, in any of the following circumstances:

- Introduction of troops into actual hostilities.
- Introduction of troops, equipped for combat, into a foreign country.
- Greatly enlarging the number of troops, equipped for combat, in a foreign country.

B-7. The President is required to make such reports within 48 hours of the triggering event, detailing—

- The circumstances necessitating introduction or enlargement of troops.
- The constitutional or legislative authority upon which he bases his action.
- The estimated scope and duration of the deployment or combat action.

B-8. Since the War Powers Resolution was passed over the veto of President Nixon, no President has either conceded its constitutionality or complied fully with its mandates.

International Law and Treaties

B-9. The U.N. Charter became effective on 24 October 1945 after being ratified by the United States and a majority of other signatories. The U.N. Charter mandates that all member states resolve their international disputes peacefully. It requires that member states refrain in their international relations from the threat or use of force. The U.N. Charter also provides that all nations have the right to use self-defense to combat acts of aggression against them until such time as the Security Council takes action.

B-10. The U.N. Charter provides the essential framework of authority for the use of force, effectively defining the foundations for a modern *jus ad bellum.* Inherent in its principles are the requirements for necessity, proportionality, and timeliness. Necessity involves considering—

- The exhaustion or ineffectiveness of peaceful means of resolution.
- The nature of coercion applied by the aggressor State.
- The objectives of each party.
- The likelihood of effective community intervention.

B-11. Proportionality refers to limiting force in magnitude, scope, and duration to that which is reasonably necessary to counter a threat or attack. Timeliness refers to situations in which the delay of a response to an attack or the threat of attack attenuates the immediacy of the threat and the necessity to use force in self-defense.

B-12. U.S. forces obey the law of war. The law of war is a body of international treaties and customs, recognized by the United States as binding. It regulates the conduct of hostilities and protects noncombatants. The main law-of-war protections come from The Hague and Geneva Conventions. They apply at the tactical and operational levels and are summarized in ten rules shown in table B-1.

Table B-1. Summary of the law-of-war rules

• Fight only enemy combatants.
• Do not harm enemies who surrender; disarm them and turn them over to the chain of command.
• Do not kill or torture detainees.
• Collect and care for the wounded, whether friend or foe.
• Do not attack medical personnel, facilities, or equipment.
• Destroy no more than the mission requires.
• Treat all civilians humanely.
• Do not steal; respect private property and possessions.
• Do one's best to prevent violations of the law of war.
• Report all violations of the law of war to superiors.

B-13. During security force assistance operations, commanders must be aware of Common Article 3 of the Geneva Conventions and the status of insurgents under the laws of the host nation. One article contained in all four of the Geneva Conventions—Common Article 3—is specifically intended to apply to internal armed conflicts:

> *In the case of armed conflict not of an international character occurring in the territory of one of the High Contracting Parties, each Party to the conflict shall be bound to apply, as a minimum, the following provisions:*
>
> *(1) Persons taking no active part in the hostilities, including members of armed forces who have laid down their arms and those placed "hors de combat" by sickness, wounds, detention, or any other cause, shall in all circumstances be treated humanely, without any adverse distinction founded on race, colour, religion or faith, sex, birth or wealth, or any other similar criteria. To this end, the following acts are and shall remain prohibited at any time and in any place whatsoever with respect to the above-mentioned persons: (a) Violence to life and person, in particular murder of all kinds, mutilation, cruel treatment and torture; (b) Taking of hostages; (c) Outrages upon personal dignity, in particular humiliating and degrading treatment; (d) The passing of sentences and the carrying out of executions without previous judgment pronounced by a regularly constituted court, affording all the judicial guarantees which are recognized as indispensable by civilized peoples.*
>
> *(2) The wounded and sick shall be collected and cared for. An impartial humanitarian body, such as the International Committee of the Red Cross, may offer its services to the Parties to the conflict. The Parties to the conflict should further endeavor to bring into force, by means of special agreements, all or part of the other provisions of the present Convention. The application of the preceding provisions shall not affect the legal status of the Parties to the Conflict.*

HOST-NATION LAW AND STATUS-OF-FORCES AGREEMENTS

B-14. After considering the type of baseline protections represented by fundamental human rights law, the military leader must be advised in regard to the other bodies of law that leader should integrate into planning and execution. This includes consideration of host-nation law. Because of the nature of most noninternational armed conflict missions, commanders and staffs must understand the technical and pragmatic significance of host-nation law within the area of operations. Although in theory, understanding the application of host-nation law during military operations is perhaps the simplest component, in practice it is perhaps the most difficult.

B-15. Judge advocates must recognize the difference between understanding the technical applicability of host-nation law and that law's application to control the conduct of U.S. forces during operations. In short, the significance of host-nation law declines in proportion to the movement of the operation toward violent conflict. Judge advocates should understand that U.S. forces enter other nations with a legal status that exists anywhere along a notional legal spectrum.

B-16. Judge advocates and planners must be familiar with any status-of-forces agreements or status-of-mission agreements that may be applicable. In any given country, there may be agreements short of status-of-forces agreements, such as diplomatic notes, on point. Relevant international documents affecting military operations may be difficult to locate. Several sources are available for locating applicable international agreements governing the status of U.S. forces or affecting military operations. Department of State (DOS) publications, such as *Treaties in Force,* contain unclassified international agreements. Both the relevant geographic combatant command's legal office and the defense attaché or military assistance group at the embassy normally have access to host-nation or international agreements impacting a military operation.

B-17. Status-of-forces and other forms of agreements frequently exist. They are essentially contractual agreements or treaties between two or more nations that establish the legal status of military personnel in foreign countries. Topics that are usually covered in a status-of-forces agreement include criminal and civil jurisdiction, taxation, and claims for damages and injuries. In the absence of an agreement or some other arrangement with the host nation, DOD personnel in foreign countries may be subject to its laws. Commanders ensure that all personnel understand status of U.S. forces in the area of operations and are trained accordingly.

LEGAL CONSTRAINTS ON THE SECURITY FORCE ASSISTANCE MISSIONS

B-18. U.S. law and regulation play a key role in establishing the parameters by which military forces may conduct security force assistance missions. These parameters tend to constitute constraints on the activities of military units. They range from the rules of engagement in combat situations to the ability to spend government funds in furtherance of a training or support mission.

GENERAL PROHIBITION ON ASSISTANCE TO POLICE

B-19. Historically, DOD is not the lead government department for assisting foreign governments. DOS is the lead when U.S. forces provide security assistance—military training, equipment, and defense articles and services—to host-nation governments. The Foreign Assistance Act of 1961 specifically prohibits assistance to foreign police forces except within specific exceptions and under a Presidential directive. When providing assistance to training, DOS provides the lead role in police assistance through its Bureau of International Narcotics and Law Enforcement Affairs. The President, however, may delegate this role to other agencies. This was done in 2004, when the President signed a decision directive granting the commander, United States Central Command, authority to train and equip Iraqi police.

TRAINING AND EQUIPPING FOREIGN SECURITY FORCES

B-20. All training and equipping of foreign security forces are specifically authorized. U.S. laws require Congress to authorize expenditures for training and equipping foreign forces. The laws of the United States also require DOS to verify that the host nation receiving the assistance is not in violation of human rights. The Secretary of Defense may authorize deployed U.S. forces to train or advise host-nation security forces as part of the mission. In this case, DOD personnel, operations, and maintenance appropriations provide an incidental benefit to those security forces. Numerous other programs to assist foreign forces are paid for with funds appropriated by Congress for that purpose. Consultation with a staff judge advocate or legal advisor early in the planning ensures the availability of funds for missions to train and equip foreign forces.

RULES OF ENGAGEMENT

B-21. *Rules of engagement* are directives issued by competent military authority that delineate the circumstances and limitations under which United States forces will initiate and/or continue combat engagement with other forces encountered (JP 1-02). Often these directives are specific to the operation. If there are no operation-specific rules of engagement, U.S. forces apply standing rules of engagement. When operating with a multinational force, commanders must coordinate the rules of engagement thoroughly.

FISCAL LAW CONSIDERATIONS

B-22. In security force assistance missions, like all operations, commanders require specific authority to expend funds normally found in the annual DOD appropriations act. As a general rule, operation and maintenance funds may not be used for security force assistance missions. In recent years, Congress appropriated additional funds to commanders for conducting more complex stability operations that were not typically covered by operation and maintenance funds. Examples include the Commander's Emergency Response Program (CERP), the Iraq Relief and Reconstruction Fund, Iraq Freedom Fund, and Commander's Humanitarian Relief and Reconstruction.

B-23. DOS has the primary responsibility, authority, and funding to conduct foreign assistance on behalf of the U.S. government. Foreign assistance encompasses any and all assistance to a foreign nation, including security assistance (assistance to the internal police forces and military forces of the foreign nation), development assistance (assistance to the foreign government in projects that will assist the development of the foreign economy or their political institutions), and humanitarian assistance (direct assistance to the population of a foreign nation). The legal authority for DOS to conduct foreign assistance is found in the Foreign Assistance Act of 1961.

B-24. There are two exceptions to the general rule requiring the use of Title 22 funds for foreign assistance. The first exception is for interoperability, safety, and familiarization training. DOD may fund the training (as opposed to goods and services) of foreign militaries with operation and maintenance dollars only when the purpose of the training is to enhance interoperability, familiarization, and safety training. Operation and maintenance funds may not be used for security force assistance training. This exception applies only to interoperability training. The second exception is congressional appropriation or authorization to conduct security foreign assistance. This exception allows DOD to fund foreign assistance operations if Congress has provided a specific appropriation or authorization to execute the mission.

Constraints on Funding

B-25. The Foreign Operations Appropriation Act (an annual DOS appropriations act) was enacted in 1997. Known as the Leahy Amendment, it prohibits the U.S. government from providing funds to a unit of the security forces of a foreign country if DOS has credible evidence that the unit has committed gross violations of human rights. The Leahy Amendment contains additional constraints on government funding of security force assistance missions. The law restricts funding until the Secretary of State determines and reports that the government of such country is taking effective measures to bring the responsible members of the security forces unit to justice.

B-26. Congress specifically appropriates funds for foreign assistance. The United States Agency for International Development (USAID) expends such funds under the legal authorities in Title 22. Provisions of Title 10 may also authorize amounts of money for these purposes. Standing funding authorities are narrowly defined and generally require advance coordination within DOD and DOS.

B-27. Effective foreign forces need training and equipment. U.S. laws require Congress to authorize such expenditures. U.S. laws also require DOS to verify that the host nation receiving the assistance is not in violation of human rights.

Funding Authorization for Security Force Assistance

B-28. All training and equipping of foreign security forces must be specifically authorized. Military and civilian personnel, operations, and maintenance appropriations should typically provide only an incidental benefit to those security forces. All other weapons, training, equipment, logistic support, supplies, and services provided to foreign forces must be paid for with funds appropriated by Congress for that purpose. Examples include the Iraq Security Forces Fund and the Afghan Security Forces Fund. Moreover, the President must give specific authority to DOD for its role in training and equipping foreign security forces. In May of 2004, the President signed National Security Presidential Directive (NSPD) 36 that made the commander, U.S. Central Command, under policy guidance from the chief of mission, responsible for coordinating all U.S. government efforts to organize, train, and equip Iraqi Security Forces, including police. Absent such a directive, DOD lacks authority to take the lead in assisting a host nation to train and equip its security forces.

Key Security Assistance and Foreign Assistance Funding Programs

B-29. Congress has appropriated funds to allow DOS to conduct its foreign assistance mission, including the following activities:

- Foreign Military Financing Program.
- International Military Education and Training Program.
- Economic Support Fund.
- Peacekeeping operations.
- Antiterrorism assistance.
- Global humanitarian demining.

- Refugee assistance.
- Personnel details.

B-30. DOS administers the following programs but no funds are appropriated to sustain them:

- Foreign Military Sales Program.
- Foreign Military Lease Program.
- Economy Act Security Assistance.
- U.S. Government Commodities and Services Program.
- Direct Commercial Sales Program.

B-31. Additional special support created by law for the foreign assistance mission includes—

- Excess defense articles.
- Presidential drawdowns.

B-32. DOS directly, or indirectly through USAID, finances numerous development assistance activities to address the following needs:

- Rule of law.
- Governance.
- Agriculture and nutrition.
- Population control.
- Health.
- Education.
- Energy.
- Environment improvement.

B-33. CERP (see paragraph B-19) is an example of a targeted humanitarian assistance fund program. CERP's primary purpose is to enable military commanders to respond to urgent humanitarian relief and reconstruction requirements within their fields of interest by carrying out programs that will immediately assist the host-nation populace. CERP was originally funded with seized Iraqi assets, but Congress later appropriated U.S. funds for the purpose. CERP is not applicable to missions outside Iraq and Afghanistan. Future missions, though, may have similar funding sources established to facilitate a humanitarian assistance mission. Commanders consult with the servicing judge advocate to determine the availability of funds.

Appendix C
Information Engagement

INFORMATION SUPERIORITY

C-1. Information shapes a commander's operational environment at every level. Twenty-first century conflicts occur in operational environments with instant communication. Because information systems are everywhere and exposure to news and opinion media is pervasive, an individual's words and actions can have immediate operational and strategic implications. Gaining and maintaining the trust of key publics has never been more important. Army forces conduct operations in the public eye and must maintain public confidence by being consistent in word and deed.

C-2. *Information superiority* is the operational advantage derived from the ability to collect, process, and disseminate an uninterrupted flow of information while exploiting or denying an adversary's ability to do the same (JP 3-13). Advisors build partner capacity in this mission area. Advisors ensure that host-nation counterparts integrate information superiority into operations as carefully as fires, maneuver, protection, and sustainment. Army information tasks contribute to information superiority. Information tasks are used to shape the operational environment (see FM 3-0). The information tasks are—

- Information engagement.
- Command and control warfare.
- Information protection.
- Operations security.
- Military deception.

C-3. Advisors prioritize the five information tasks based on the host nation's capabilities and strengths. This appendix discusses some information engagement capabilities that advisors can expect to employ while conducting security force assistance, as well as considerations when working with the media. For a further discussion of the other information tasks, see FM 3-0.

INFORMATION ENGAGEMENT CAPABILITIES

C-4. *Information engagement* is the integrated employment of public affairs to inform U.S. and friendly audiences; psychological operations, combat camera, and U.S. government strategic communications and defense support to public diplomacy, and other means necessary to influence foreign audiences; and leader and Soldier engagements to support both efforts (FM 3-0).

C-5. In general, advisors focus on information engagement because land operations occur among civilian populations. Advisors can have a significant impact by focusing on this task. Advisors work with host-nation counterparts to communicate information to the populace, build their trust and confidence in friendly forces, promote popular support for friendly operations, and positively influence public perceptions and behavior. Advisors integrate information engagement throughout the operations process.

C-6. Advisors ensure host-nation counterparts focus their information engagement activities on achieving desired effects locally. However, land operations always occur in a regional and even global context. Therefore, local information engagement activities must support and complement higher headquarters information engagement activities, host-nation and U.S. government strategic communications guidance (when available), and broader host-nation and U.S. government policy where applicable.

C-7. Three information engagement capabilities deserve particular attention as the advisor can directly influence them. These are public affairs, combat camera, and leader and Soldier engagements.

Public Affairs

C-8. First, public affairs can have a profound and lasting effect on public perceptions of operational success or failure. This effect may have strategic consequences, even when related to tactical operations. Therefore, advisors ensure that public affairs efforts are integrated at tactical, operational, and strategic levels. (See paragraphs C-10 through C-20 for more information to help advisors working directly with the media.)

Combat Camera

C-9. Second, combat camera is used to document a wide range of tactical and operational successes and to counter enemy propaganda. If there is no dedicated host-nation combat-camera capability, the advisor recommends that a host-nation soldier be designated, equipped, and trained for this function. If that is not possible, the advisor should obtain a digital camera and, if necessary, photograph relevant host-nation activities personally. Positive photographs can have a tremendous benefit, and digital photos are easily taken and transmitted.

Leader and Soldier Engagements

C-10. Third, leader and Soldier engagements are a fundamental and highly effective method to influence people. They involve face-to-face interaction by leaders and Soldiers to influence the perceptions of the local populace. Military leaders engage key communicators and civilian leaders (often referred to as key leaders) through carefully planned meetings. Examples of key leaders include local tribal elders, elected or appointed officials, military and police commanders, and nongovernmental organization leaders.

WORKING WITH THE MEDIA

C-11. The appropriate public affairs officer normally manages media contacts. However, this is not always possible. A deployed advisor team seldom has a public affairs officer representative.

C-12. Occasions may arise when the host-nation press wants to question advisors. Even if an advisor team lacks a public affairs officer representative, advisors must avoid refusing to speak to the media. Apparent refusal to speak with accredited members of the media, for whatever reason, may inadvertently create strong negative impressions with strategic implications. Therefore, advisor teams designate a media spokesperson and keep this individual well informed. This individual coordinates media contacts for the advisor team. This individual ensures that official, published public affairs officer guidance and command policies regarding media relations, including anonymity and operations security, are observed. The following paragraphs provide general media relations guidance for advisor teams.

Enhancing Relationships with the Media

C-13. Advisors strive to convey a positive impression and enhance relationships between the media and the military. If deemed acceptable within the public affairs guidance and command policies, the advisor team's designated media spokesperson—

- Maintains a list of reliable reporters and editors that cover operations in the team's operational area and keeps them informed of significant activities.
- Answers media inquiries promptly, accurately, and courteously.
- Finds out reporters' deadlines and uses them to the friendly force's advantage.
- Stresses the human aspects of a story that readers, viewers, or listeners can identify with, including the negative impact of opposing operations on the people.
- Points out how advisors and host-nation counterparts are working to address the needs of the unfortunate.
- Encourages the media to observe constructive work the host nation and advisors are doing and to interview them about their efforts.
- Identifies a location in advance for the press to take photos and videos.

Correcting Public Misconceptions

C-14. If information is withheld, journalists often fall back on speculation. Such speculation is usually inaccurate. However, it is often near enough to the truth that it is easily accepted by large sections of the public and possibly by established governments. In addition, partisan sources may find it advantageous to leak part of a story to the press to build up public support for their position. On occasion, such activities can grow into fully orchestrated press campaigns.

C-15. With modern satellites and communications technology, media are able to distribute reports and photographs extremely quickly. Incidents, sometimes fabricated or slanted toward a partisan viewpoint, may be aired in living rooms across the globe the same day.

C-16. However, advisors and host-nation counterparts frequently can correct public misconceptions and provide accuracy, adequate context, and proper characterization promptly. They can turn a potential liability into a major advantage by remaining prepared to coordinate and communicate with the media. They can take advantage of the media's powerful and far-reaching ability to communicate with critical audiences.

C-17. If the media report inaccurate or damaging information, advisors use caution in deciding on a course of action. Advisors (usually through the team's designated media spokesperson) inform their chain of command. Advisors consider the following questions:

- Is it important enough to correct or is it a detail that—in the long term—is not really important?
- How damaging is the charge, criticism, or error? Will a correction simply give greater visibility to an unfavorable point of view?
- Is a correction worth a restatement of the entire problem to new audiences who did not read or see it the first time?
- Is it possible to target only the audience originally exposed to the story?
- Can significant gain result from pointing out an error?

C-18. When incorrect or misleading information is released to the public through host-nation government or media sources, the advisor team determines a course of action appropriate to the situation. It must include a plan to ensure that relevant host-nation government officials and media recognize the inaccuracy of the information. This may prevent future incidents. Operational and combat experience since 2001 shows that in the struggle for information superiority, friendly forces must be candid and honest. Additionally, they must not allow lies disseminated by enemies to stand unchallenged, particularly in areas of great international interest.

Interviews With Advisor Team Members

C-19. Advisor teams sometimes arrange for media interviews with selected advisor team members. Properly conducted interviews can support the mission and even improve morale. When preparing for an interview, the team's designated spokesperson—

- Provides all team members with a single and simple theme to convey to the press, sometimes in the form of a mission narrative (a succinct account of the mission as a whole, explained consistently with the host nation's cultural view).
- Conducts rehearsals in front of video cameras, using mock interviews to practice communication of the theme.
- Selects interviewees based on their understanding of the theme covered, not their comfort level in front of a reporter. (Shyness is a normal trait that may add a human touch to a sensitive situation.)

C-20. If an advisor team member does not know the answer to a reporter's question, the individual should attempt to find out and get back to the reporter or refer the reporter to an appropriate source. This may establish the team as a helpful information source and develop a relationship for future balanced coverage.

C-21. Communicating with the media is the most difficult aspect of information engagement. Anything advisors say or do publicly can impact the mission. Therefore, advisors ensure consistency with national policy, professionalism, and clear communication. Advisors must know and follow the policies of the geographic combatant commander, the American Embassy, and the appropriate public affairs officer regarding media interviews. Additionally, advisors follow the specific guidelines shown in table C-1.

Table C-1. Specific guidelines for media interviews

- Maintain a friendly, yet professional attitude.
- Remain positive, greet interviewers, and welcome questions. (This is difficult under stressful or tragic conditions; however, a calm, mature appearance earns respect from audiences.)
- Use language that is clear and easy to understand.
- Avoid military jargon or terminology that others can misinterpret.
- Never enter an interview unprepared.
- Prepare notes and study them carefully; use facts to back up statements.
- Ensure the facts supporting their position are up to date and come to mind easily.
- Anticipate possible questions and think about various responses.
- Rehearse before interviews on location if possible and, if time permits, role-play tough questions.
- Learn and use reporters' names.
- Research interviewers' organizations, including—
  - The organizations' views or views they defend.
  - Their underlying motivations.
  - How members of the organizations previously conducted interviews.
- Keep the interests of the local nationals or other beneficiaries of the mission at the forefront.
- Avoid talking from the point of view of U.S. interests.
- Finish answering one question before answering another if two or more questions are asked at the same time.
- Avoid speculation and give only factual information that can be verified.
- Use caution in quoting statistics as such data may be easily disputed or reinterpreted.
- Avoid repeating a question that contains incorrect information or inflammatory language (this may result in a misquotation).
- Strive to tell the truth.
- Avoid saying *no comment*. (It gives the appearance the advisor has something to hide.)
- If unsure of something, admit it rather than mistakenly lie.
- If questioned about another agency's activities, refer to that agency for comment.
- Do not speak for other organizations.
- Avoid exaggeration or claims that cannot be backed up.
- Refuse all off-the-record discussions.
- Expect any statement they make to be quoted publicly.
- Avoid criticizing other groups or organizations, and avoid committing information fratricide
- Remind reporters that photographing recognizable dead Soldiers, charts, maps, supply depots, or electronic warfare assets is off limits.
- Remind reporters that their personal security is not a primary military concern.
- Remain objective and calm.
- Demonstrate conduct befitting a representative the U.S. government
- Never give personal opinions and beliefs.
- Avoid reacting emotionally when interviewers appear skeptical or hostile.
- Maintain focus on the mission.

Glossary

SECTION I – ACRONYMS AND ABBREVIATIONS

ABCA	American, British, Canadian, Australian, and New Zealand Armies Program
AR	Army regulation
BCT	brigade combat team
CCIR	commander's critical information requirement
CERP	Commander's Emergency Response Program
COA	course of action
COIST	company intelligence support team
DA	Department of the Army
DOD	Department of Defense
DODD	Department of Defense directive
DOS	Department of State
DOTMLPF	doctrine, organization, training, materiel, leadership and education, personnel, facilities
FM	field manual
FSF	foreign security forces
IDAD	internal defense and development
ISR	intelligence, surveillance, reconnaissance
JP	joint publication
METT-TC	mission, enemy, terrain and weather, troops and support available, time available, civil considerations (mission variables)
MTT	military transition team
NSPD	national security Presidential directive
PIR	priority intelligence requirement
S-2	intelligence staff officer
S-3	operations staff officer
S-4	logistics staff officer
SFA	security force assistance
TC	training circular
UN	United Nations
US	United States
USAID	United States Agency for International Development

SECTION II – TERMS

foreign internal defense

(joint) Participation by civilian and military agencies of a government in any of the action programs taken by another government to free and protect its society from subversion, lawlessness, and insurgency. (JP 3-05)

foreign military sales

(joint) That portion of United States security assistance authorized by the Foreign Assistance Act of 1961, as amended, and the Arms Export Control Act of 1976, as amended. This assistance differs from the Military Assistance Program and the International Military Education and Training Program in that the recipient provides reimbursement for defense articles and services transferred. (JP 1-02)

***foreign security forces**

Forces, including but not limited to military, paramilitary, police, and intelligence forces; border police, coast guard, and customs officials; and prison guards and correctional personnel, that provide security for a host nation and its relevant population or support a regional security organization's mission.

host nation

(joint) A nation that receives the forces and/or supplies of allied nations, coalition partners, and/or NATO organizations to be located on, to operate in, or to transit through its territory. (JP 3-57)

humanitarian and civic assistance

(joint) Assistance to the local populace provided by predominantly US forces in conjunction with military operations and exercises. This assistance is specifically authorized by Title 10, United States Code, section 401, and funded under separate authorities. Assistance provided under these provisions is limited to (1) medical, dental, and veterinary care provided in rural areas of a country; (2) construction of rudimentary surface transportation systems; (3) well drilling and construction of basic sanitation facilities; and (4) rudimentary construction and repair of public facilities. Assistance must fulfill unit training requirements that incidentally create humanitarian benefit to the local populace. (JP 3-57)

internal defense and development

(joint) The full range of measures taken by a nation to promote its growth and to protect itself from subversion, lawlessness, and insurgency. It focuses on building viable institutions (political, economic, social, and military) that respond to the needs of society. (JP 3-07.1)

international military education and training

(joint) Formal or informal instruction provided to foreign military students, units, and forces on a nonreimbursable (grant) basis by offices or employees of the United States, contract technicians, and contractors. Instruction may include correspondence courses; technical, educational, or informational publications; and media of all kinds. (JP 1-02)

security assistance

(joint) Group of programs authorized by the Foreign Assistance Act of 1961, as amended, and the Arms Export Control Act of 1976, as amended, or other related statutes by which the United States provides defense articles, military training, and other defense-related services, by grant, loan, credit, or cash sales in furtherance of national policies and objectives. (JP 3-57)

security force assistance

The unified action to generate, employ, and sustain local, host-nation, or regional security forces in support of legitimate authority. (FM 3-07)

security sector reform

The set of policies, plans, programs, and activities that a government undertakes to improve the way it provides safety, security, and justice. (FM 3-07)

References

Field manuals and selected joint publications are listed by new number followed by old number.

REQUIRED PUBLICATIONS

These documents must be available to intended users of this publication.

FM 1-02 (101-5-1). *Operational Terms and Graphics.* 21 September 2004.

JP 1-02. *Department of Defense Dictionary of Military and Associated Terms.* 12 April 2001.

RELATED PUBLICATIONS

These documents contain relevant supplemental information.

JOINT AND DEPARTMENT OF DEFENSE PUBLICATIONS

Most joint publications are available online: <http://www.dtic.mil/doctrine/jpcapstonepubs.htm.>

DOD 5105.38-M. *Security Assistance Management Manual.* 3 October 2003. <http://www.dsca.osd.mil/samm/>

DODD 3000.05. *Military Support for Stability, Security, Transition, and Reconstruction (SSTR) Operations.* 28 November 2005. <http://www.dtic.mil/whs/directives/corres/html/300005.htm>

JP 3-0. *Joint Operations.* 17 September 2006.

JP 3-05. *Doctrine for Joint Special Operations.* 17 December 2003.

JP 3-07.1. *Joint Tactics, Techniques, and Procedures for Foreign Internal Defense (FID).* 30 April 2004.

JP 3-07.4. *Joint Counterdrug Operations.* 13 June 2007.

JP 3-57. *Civil-Military Operations.* 8 July 2008.

ARMY PUBLICATIONS

Most Army doctrinal publications are available online: <http://www.army.mil/usapa/doctrine/Active_FM.html>. Army regulations are produced only in electronic media. Most are available online: < http://www.army.mil/usapa/epubs/index.html>.

AR 11-31. *Army International Security Cooperation Policy.* 24 October 2007.

AR 12-1. *Security Assistance, International Logistics, Training, and Technical Assistance Support Policy and Responsibilities.* 24 January 2000.

FM 3-0. *Operations.* 27 February 2008.

FM 3-05.137. *Army Special Operations Forces Foreign Internal Defense.* 30 June 2008.

FM 3-05.202 (FM 31-20-3). *Special Forces Foreign Internal Defense Operations.* 2 February 2007.

FM 3-07. *Stability Operations.* 6 October 2008.

FM 3-24. *Counterinsurgency.* 15 December 2006.

FM 3-34.400 (FM 5-104). *General Engineering.* 9 December 2008.

FM 4-0 (100-10). *Combat Service Support.* 29 August 2003.

FM 4-02 (FM 8-10). *Force Health Protection in a Global Environment.* 13 February 2003.

FM 6-0. *Mission Command: Command and Control of Army Forces.* 11 August 2003.

FM 6-22 (22-100). *Army Leadership.* 12 October 2006.

FM 7-0. *Training for Full Spectrum Operations.* 12 December 2008.

FM 7-15. *The Army Universal Task List.* 27 February 2009.

TC 31-73. *Special Forces Advisor Guide.* 2 July 2008.

OTHER PUBLICATIONS

National Defense Strategy of 2008. Washington, DC: U.S. Government Printing Office, 2008. <www.defenselink.mil/news/2008 National Defense Strategy.pdf>

National Military Strategy of the United States of America. Washington, DC: U.S. Government Printing Office, 2004. <www.defenselink.mil/news/Mar2005/d20050318nms.pdf>

National Security Strategy of 2006. Washington, DC: U.S. Government Printing Office, 2006. <www.marforres.usmc.mil/docs/nss2006.pdf>

Quadrennial Defense Review of 2006. Washington, DC: U.S. Government Printing Office, 2006. <www.defenselink.mil/qdr/report/Report20060203.pdf>

Armed Forces. Title 10, U.S. Code. <http://www.loc.gov/law/help/guide/federal/uscode.php>

Arms Export Control Act of 1976. <http://www.loc.gov/law/help/guide.php>

The United States Constitution. <http://www.house.gov/house/Constitution/Constitution.html>

Foreign Relations and Intercourse. Title 22, U.S. Code. <http://www.loc.gov/law/help/guide/federal/uscode.php>

Foreign Operations Appropriation Act of 1997. <http://www.loc.gov/law/help/guide.php>

Foreign Assistance Act of 1961. <http://www.loc.gov/law/help/guide.php>

The Geneva Conventions of 1949. <http://www.loc.gov/rr/frd/Military_Law/MLR_site-map.html>

National Security Presidential Directive 36. "United States Government Operations in Iraq." 11 May 2004.

Treaties in Force 2007. <http://www.state.gov/s/l/treaty/treaties/2007>

Charter of the United Nations. <http://www.un.org/aboutun/charter/>

War Powers Act of 1973. <http://www.thecre.com/fedlaw/legal22/warpow.htm.>

OUTSIDE RECOMMENDED READING

von Steuben, Friedrick Wilhelm. *Regulations for the Order and Discipline of the Troops of the United States.* Philadelphia: Styner and Cist, 1779.

PRESCRIBED FORMS

None

REFERENCED FORMS

DA Form 2028. Recommended Changes to Publications and Blank Forms

Index

Entries are by paragraph number.

Entries are by paragraph number.

Entries are by paragraph number.

Entries are by paragraph number.

Made in the USA
Monee, IL
05 November 2026

67810962R00164